RECLAIMING EDEN

RECLAIMING EDEN

Responsible Living, Engineering, and Architectures

edited by

David S.-K. Ting

Jacqueline A. Stagner

JENNY STANFORD
PUBLISHING

Published by

Jenny Stanford Publishing Pte. Ltd.
101 Thomson Road
#06-01, United Square
Singapore 307591

Email: editorial@jennystanford.com
Web: www.jennystanford.com

British Library Cataloguing-in-Publication Data
A catalogue record for this book is available from the British Library.

Reclaiming Eden: Responsible Living, Engineering, and Architectures

ISBN 978-981-5129-31-1 (Hardcover)
ISBN 978-1-003-49612-0 (eBook)

Dedication

To everyone who strives to reclaim Earth to its original beauty and tranquility.

Contents

Preface

Life on Earth is both challenging and beautiful. *Reclaiming Eden* is about responsible living, engineering, and architectures, aiming to mitigate environmental deterioration by reclaiming land around the world to an ecologically sustainable stage. These endeavors will enable us to pass forward a beautiful tomorrow for our grandchildren in the long run, and our children and ourselves in the immediate future. **Ting** and **Stagner** ascertain that it makes splendid sense to reclaim Eden in Chapter 1, "Eden: It Makes Splendid Sense to Reclaim." As Earth dwellers, we have the responsibility to tend this garden. While they are not often broadcast by major media, there have been a few success stories about reclaiming Eden-like gardens in places around the world. Understanding the outstanding challenges can better position us to further the restoration on a worldwide scale. A comprehensive disclosure on living responsibly is provided in Chapter 2, "Living Responsibility: Adapting Waste Reduction to Reclaim Eden?" **Reader** rightly conveys that the Garden of Eden was initially stewarded by two unclothed ageless human custodians without any mention of waste. Not too many generations later, waste has become a serious issue to be reckoned with. Population growth is one cause, and new waste associated with renewable energy is another. Waste management and reduction should be enunciated, and efforts must be invested now before the problem goes out of control. Recycling is the second most effective means of reducing waste, after reduction. **Balo**, **Boydak**, and **Sua** furnish "Harbor-House Design with Recycled Insulation Materials for Sustainable Development," as Chapter 3. They show that proper utilization of building information modeling, along with building energy modeling, can conserve money, time, and resources, furthering sustainable and green constructions. Improving the efficiency of the many man-made systems for sustaining everyday living is also critical in Reclaiming Eden. Integrating two or more renewable energy systems can better harness the available free energy. Photovoltaic-thermoelectric hybrid power systems are such an example. Three aspects of photovoltaic-thermoelectric hybrid power systems—optical design, heat transfer, and geometric structure—are discussed in detail by **Wang**, **Henshaw**, and **Ting** in Chapter 4, "Recent Progress in Photovoltaic-Thermoelectric Hybrid Power Systems." **Choobchian**, **Ghorbannejad**, **Amghani**, and **Azadi** present "Barriers to the Deployment of Renewable Energy Technologies" in Chapter 5. The case study on the situation in Iran identifies 28 barriers. It is not surprising to see that politics and the regulatory framework are the main factors in preventing renewable energy penetration. Political will is a must if we wish to bring back Eden to Earth. Creatures such as birds are not subjected to political and regulatory hindrances when it comes to effective harnessing of renewable energy. In Chapter 6, "Solar Enhanced Soaring: Tapping into Ambient Solar Energy for Optimized Airfoil Performance," **Akhtar**, **Ting**, and **Kalathi** reveal the effect of solar heating of one surface of an ideal bird's wing represented by an NACA0012 airfoil. Detectable changes in the lift and drag coefficients can be achieved with a moderate 10°C temperature difference between the two surfaces. Thus, it is clear that, to green life on Earth, we can benefit from a better understanding of the multitude of intelligently designed creatures and creations in the natural world. Talking about imitating nature, **Jana**, **Lakkaraju**, **Stagner**, and **Ting** present "Thermal Regulation due to Vertical Greening Systems: A Review," in Chapter 7. Incorporating natural vegetation on man-made buildings can simultaneously save energy and soothe our souls. The effectiveness of vertical greening systems, however, is a challenge to deduce. Some promising performance indicators are proposed, pending future studies to solidify the evaluation standard. Wood can save mankind from many troubles, for example, the Ark that kept Noah and his companions afloat was constructed from gofer wood, according to Genesis 6:14. Instead of adhesives and metal fasteners to join pieces of wood, dovetail joinery can be employed. This is the topic of Chapter 8, "Evolutionary Structurally-Sound-Design-Process of Dovetail Massive Wooden Board Elements as a Sustainable and Environmentally Friendly Solution," authored by **Ilgın**, **Karjalainen**, and **Pajunen**. The integration of dovetail-based components in non-glued and non-metal wooden boards is a sustainable path to advance tomorrow's architectures. While it is well suited for buildings, wood is not the best fuel for cooking because of emission issues, even though, as a biofuel, it is carbon neutral. **Abutu**, **Oruonye**, **Abubakar**, and **Auwal** promote the use of electricity for cooking in Chapter 9, "Performance Assessment of Electric-Powered Food Roasting Machine." They constructed an electric powered food roasting machine using local materials, mostly scrap, in Jalingo, Nigeria. The ease of control allows precise control of the cooking process, which varies depending on the amount and type (moisture content) of the food. There can be no Garden without soul-soothing waters. This volume concludes with a highlight of improvements to a coral reef, "Coral Reef Ecosystem Enhancement in Singapore's Highly Urbanised Port," as Chapter 10 by **Ng**, **Sam**, **Kikuzawa**, **Toh**, and **Chou**. The densely populated city-state of Singapore proves that we can keep our ecosystem intact when urbanization occurs. Yes, Reclaiming Eden is realizable.

Acknowledgments

We are indebted to the experts who exerted considerable effort in compiling the chapters that make up this volume. The anonymous reviewers ensured each manuscript is up to standard. The editors continue to savor the pleasant relationship with an amiable publisher, Jenny Rompas. This impossible undertaking is made possible only by timely providence from Above.

David S.-K. Ting and Jacqueline A. Stagner
Turbulence & Energy Laboratory
University of Windsor

Chapter 1

Eden: It Makes Splendid Sense to Reclaim

David S.-K. Ting and Jacqueline A. Stagner

Turbulence & Energy Laboratory, University of Windsor, Windsor, Ontario, Canada

stagner@uwindsor.ca

Aesthetic, ecologically friendly gardens like Eden are essential in overcoming the anthropogenic deterioration of our environment. This chapter provides an overview of the contents of this timely volume. It highlights some of the unsung successes in Reclaiming Eden-like gardens over the years. Outstanding challenges to spreading the Reclaiming of Eden worldwide are presented, differentiating these from many presumed obstacles that have proven to be false. To brighten tomorrow, let us strive to reclaim Eden together.

1.1 Eden and Our Responsibility

Hódosy [2023] provided one of the latest discussions concerning the meaning of Genesis 1: 28, which in the New International Version says:

> God blessed them and said to them, "Be fruitful and increase in number; fill the Earth and subdue it. Rule over the fish in the

Reclaiming Eden: Responsible Living, Engineering, and Architectures
Edited by David S.-K. Ting and Jacqueline A. Stagner

ISBN 978-981-5129-31-1 (Hardcover), 978-1-003-49612-0 (eBook)
www.jennystanford.com

sea and the birds in the sky and over every living creature that moves on the ground."

Hódosy explained the word "rule" as "humans should be the guardians of the non-human world, and it is they who should take responsibility for the corruption of the ecosystems." This is very close to The Message translation, which uses "responsible" instead of "rule," that is, "Be responsible for fish in the sea and birds in the air, for every living thing that moves on the face of Earth." Further, The Voice translation expresses our caring responsibility explicitly, "I make you trustees of My estate, so care for My creation and rule over the fish of the sea, the birds of the sky, and every creature that roams across the Earth."

1.2 Eden and Its Fall

Hódosy, among many others, speculated the magnificent and lush Garden of Eden to represent the area around the Persian Gulf, southern Mesopotamia[1] [Hódosy, 2023]. It was further argued, based on previous studies, that irresponsible anthropogenic changes have resulted in the area being not so fertile and lush today. The pre-Islamic Persia is known for its many beautiful Eden-like gardens [Fallahi et al., 2020]. Fallahi et al. [2020] highlighted Baghdad, that is, the garden of justice, and in this context, Bagh-e-Eram, or Eram Garden, or Garden of Eden. Many Persian gardens have been destroyed because of wars over the centuries.

1.3 Some Challenges in the Reclaiming of Eden

What can we do to reclaim Eden? First, we may not be able to fully realize the original Garden of Eden. Secondly, a virtuous striving such as this is expected to face steep opposition and many challenges. Be that as it may, what is not obvious is that the striving itself, let alone the realization, is beneficial on numerous fronts. For example, Spaling and Kooy [2019] expounded on conservation agriculture, specifically, Farming God's Way, which

[1]The word Mesopotamia means 'between rivers.' In this case, the rivers are Tigris and Euphrates [Richardson and Hussain, 2006].

has spread throughout Sub-Saharan Africa and beyond. It is a development intervention meant to increase crop yield and sustainability and improve food security.

The destructions many Eden-alike gardens have experienced over the centuries, such as those mentioned by Fallahi et al. [2020], mostly because of conflicts and invasions, could not stop the rebuilding of the gardens. But, how do you handle tensions between production ethics and environmental ethics of agricultural soils? An answer to this is detailed by van Mansvelt et al. [2021]. They proposed to protect soil based on the optimization of purposefulness in agricultural production and the terrestrial ecosystem. Factors considered are the right of mankind to use soil, soil's purposefulness, and soil's right to be protected, including its physical, chemical, and biological cycles. The sustainable solution requires appropriate restoration of degraded soils and water, acknowledging the rights of soils, and ensuring soil health, etc. In other words, human beings will be at a grave loss if Eden falls.

There is also a perception that ecosystem restoration will conflict with food production because of their competition for scarce water. Ward et al. [2021] performed an integrated basin-scale hydroeconomic analysis on this by linking biophysical, hydrologic, agronomic, ecologic, economic, policy, and institutional dimensions of the partially restored Mesopotamian Marches of Western Asia. They found that restored wetlands complement key dimensions of food production. An economically optimized ecosystem restoration can be accomplished with a minimal decrease in food production. Most importantly, nimble water sharing can lead to ecosystem restoration that comes with improved food security.

Another hurdle preventing Reclaiming Eden from taking off is the lack of financial incentive, that is, it requires a significant financial investment without cash profit. This does not need to be the case, even when exotic plants requiring costly tending are involved. Belousova et al. [2021] analyzed the Eden Project, the Eden Botanical Garden, in Cornwall, England. They demonstrated that a self-sufficient, architecturally appealing, energy-efficient, and economically viable Eden can be realized. As such, they illustrated the possibility of transforming the anthropogenic landscape into botanical gardening facilities.

With positive findings such as those conveyed above, it is no wonder the rebuilding of Eden is gaining ground, although rather quietly. For example, close to the actual location of the Garden of Eden are many restoration projects; see, for example, Rubenstein and Sathikh [2021]. In this particular study, the importance of including waste in the reclamation is also stressed.

1.4 The Many Fringe Benefits of Eden

Eden is far more than ecological. The many fringe benefits associated with the Garden of Eden and in Reclaiming Eden are irrefutable. Study after study has confirmed that such ecologically green settings can serve as an antidote for negative emotions, replacing them with positive emotions. This is illustrated via multiple green-blue spaces in Yoruba cities by Adedeji [2023]. The beauty in the diversity of plant communities, an essential outcome of ecological restoration, goes far beyond the garden surface. Duley et al. [2023] identified significant diversity in the soil environmental DNA even up to a depth of 7 m that they sampled.

1.5 Moving Forward

It appears that Reclaiming Eden can be realized feasibly. To move forward effectually, the viability, along with the multitude of benefits of restoring our environment into an Eden-like garden, must be communicated to the public at large. Furthermore, we must invest in persuading a change in the attitude of every earthling. For example, Toledano [2022] proposed the adoption of Paul Ricoeur's approach to business ethics, starting with environmental entrepreneurs. Hyperbolic generosity, in terms of sharing what one has received by caring for nature and others with abundant solicitude, is the sought-after motivator.

This volume promotes the Reclaiming of Eden. It integrates and enhances the efforts at various fronts for engineering a healthier tomorrow. The best way forward is for engineers, architects, farmers, health and social professionals, and others to work together. An essential step toward Reclaiming Eden, as eluded by [Rubenstein and Sathikh, 2021], is to reduce the massive amount of anthropogenic waste in terms of food, energy, and clothing [Ting and Stagner, 2023]. As such, **G. T. Reader** comprehensively addresses this core element of living responsibility in Chapter 2, "Living Responsibility: Adapting Waste Reduction to Reclaim Eden?" Reader wonders if Reclaiming Eden can be viewed as a manifestation of the United Nation's vision of Transforming our World into a global habitat with peace, prosperity, and partnership beyond security for all by 2030. For this, one fallout is waste, and its amount must be put in check. Another thought-provoking challenge is the waste associated with greening energy production. Specifically, the disposal of solar panels, electrochemical energy storage devices, and debris associated with mining non-renewable metallic commodities for fabricating renewable energy systems.

Good insulation can reduce heating requirements during the cold season, that is, waste heat reduction. **F. Balo**, **H. Boydak,** and **L. S. Sua** enlighten us regarding the responsible utilization of recycled insulation materials to realize eco-friendly, energy-efficient constructions. This information is conveyed in Chapter 3, "Harbor-House Design with Recycled Insulation Materials for Sustainable Development." The building information modeling methodology can ease the otherwise-complex process of choosing the most suitable sustainable design criteria for the specific scenario at hand.

Reducing waste alone is not enough, we need cleaner energy. To further photovoltaic energy production, we capitalize on a portion of the waste heat by incorporating a thermoelectric generator. **X. Wang**, **P. Henshaw,** and **D. S.-K. Ting** furnish, "Recent Progress in Photovoltaic-Thermoelectric Hybrid Power Systems" in Chapter 4. It is worth noting that a good idea can be implemented in a widespread manner only when it is economically viable.

There is much room in terms of replacing dirty energy with renewables. To hasten this, existing barriers need to be identified and subsequently removed. This is the discussion of Chapter 5, "Barriers to the Deployment of Renewable Energy Technologies," by **M. Ghorbannejad**, **S. Choobchian**, **M. S. Amghani,** and

H. Azadi. They identify 28 barriers and find that the main barriers are within the political and regulatory framework. It is clear that the decision-makers must be convinced first, before any meaningful Reclaiming of Eden can be accomplished.

Birds, on the other hand, are not hindered by barriers that humans face when it comes to tapping into free ambient energy. In addition to riding on solar-induced thermals, it is believed that their body pigmentation is intelligently designed to enhance their flight via selected heating and cooling. **M. S. Akhtar**, **D. S.-K. Ting**, and **J. T. Kalathi** present a numerical study of NACA0012 airfoil with its upper surface subjected to 10 K higher or lower than the ambient temperature, "Solar Enhanced Soaring: Tapping into Ambient Solar Energy for Optimized Airfoil Performance," Chapter 6. Both lift and drag are affected by the small temperature difference.

Greening the buildings that we spend much time in can save significant energy usage while soothing the souls of the occupants. As such, **S. Jana**, **R. T. Kalathi**, **J. A. Stagner,** and **D. S.-K. Ting** bring us up to date on this topic in Chapter 7, "Thermal Regulation due to Vertical Greening Systems: A Review." To reclaim Eden on a large scale, implement vertical greening systems everywhere. To better vertical greening systems, appropriate performance is a priori.

Tomorrow's urbanization must incorporate green infra-structures and resources that are involved in everyday urban living. Other than employing recycled materials for building insulation, as covered in Chapter 3, architecturally aesthetic structures can be built in a sustainable manner. Wood is such a building ingredient, as expounded in Chapter 8, "Evolutionary Structurally-Sound-Design-Process of Dovetail Massive Wooden Board Elements as a Sustainable and Environmentally Friendly Solution," by **H. E. Ilgin**, **M. Karjalainen,** and **S. Pajunen**. For more sustainable and environmentally friendly architectures, engineered wood products must be replaced with solid and pure wood.

Electric power for cooking makes sense when the source is renewable and if it replaces the dirtier and less reliable energy source of an open fire. This is the topic of Chapter 9, "Performance Assessment of Electric-Powered Food Roasting Machine," by

J. Abutu, O. E. Daniel, M. Abubakar, and **A. Auwal**. The case study examines the performance of electric-powered ovens developed using locally available materials. The cooking time for local stable food, maize, and chicken, is evaluated at different temperatures.

The volume concludes with an important reminder for us to take good care of the water that makes Earth habitable. It is comforting to learn about the very successful "Reclaiming Eden" project in Singapore's port. Specifically, **C. S. L. Ng**, **S. Q. Sam**, **K. Y. Preslie**, **T. C. Toh**, and **L. M. Chou** present "Coral Reef Ecosystem Enhancement in Singapore's Highly Urbanized Port" in Chapter 10. Strategies to conserve and restore coral reefs are discussed in detail based on a five-year program on Sultan Shoal. Using a customized nursery, they study the transplantation of 30 coral species.

These disclosures shed much hope and assert that much more can be and should be done. The strivings expounded in the Reclaiming Eden will certainly turn the prevailing negativity into hope for a better tomorrow. To do so efficaciously, we need to bring our fellow humankind together. Helen Adams Keller taught us how to turn such a wonderful dream into reality when she pragmatically uttered, "Alone we can do so little; together we can do so much." Wangari Maathai asserted that "We can work together for a better world with man and women of goodwill, those who radiate the intrinsic goodness of humankind." Let us strive to brighten tomorrow together as we march forward Reclaiming Eden for everyone to savor.

References

J. A. Adedeji, Chapter 2, "Green-blue spaces in Yoruba cities – ecosystem services ethnography," in Ecological Urbanism of Yoruba Cities in Nigeria, pp. 43–88, *Cities and Nature*, Springer, Cham, 2023.

O. Belousova, T. Medvedeva, Z. Aksenova, "A botanical gardening facility as a method of reclamation and integration of devastated territories (based on the example of the Eden project)," *Civil Engineering and Architecture*, 9(5): 1309–1317, 2021.

E. Duley, A. Iribar, C. Bisson, J. Chave, J. Donald, "Soil environmental DNA metabarcoding can quantify local plant diversity for biomonitoring across varied environments," *Restoration Ecology*, 31(4): e13831, 2023.

E. Fallahi, P. Fallahi, S. Mahdavi, "Ancient urban gardens of Persia: concept, history, and influence on other world gardens," *HortTechnology*, 30(1): 6–12, 2020.

A. Hódosy, "Aronofsky's two visions of the ecological fall," *Hungarian Journal of English and American Studies*, 29(1): 78–102, 2023.

C. J. Richardson, N. A. Hussain, "Restoring the Garden of Eden: an ecological assessment of the marches of Iraq," *Bioscience*, 56(6): 477–489, 2006.

M. Rubenstein, P. Sathikh, "Eden in Iraq: a wastewater design project as bio-art – a confluence of nature and culture, design and ecology, in Southern Iraq marshes," *AI & Society*, 36: 1377–1388, 2021.

H. Spaling, K. V. Kooy, "Farming God's Way: agronomy and faith contested," *Agriculture and Human Values*, 36: 411–426, 2019.

T. Z. T. Ting, J. A. Stagner, "Fast fashion – wearing out the planet," *International Journal of Environmental Studies*, 80(4): 856–866, 2023.

N. Toledano, "Restoring the Garden of Eden: A Ricoeurian view of the ethics of environmental entrepreneurship," *Business Ethics, the Environment & Responsibility*, 31: 1174–1184, 2022.

J. D. van Mansvelt, P. C. Struik, A. Bos, W. Daub, D. Sprangers, M. van den Berg, M. Vingerhoets, K. Zoeteman, "Changing ground: handling tensions between production ethics and environmental ethics of agriculture soils," *Sustainability*, 13: 13291, 2021.

F. A. Ward, D. Salman, S. A. Amer, "Managing food – ecosystem synergies to sustain water resource systems," *Science of the Total Environment*, 796: 48945, 2021.

Chapter 2

Living Responsibility: Adapting Waste Reduction to Reclaim Eden?[1]

Graham T. Reader

Department of Mechanical, Automotive and Materials Engineering, University of Windsor, Windsor, Ontario, Canada

greader@uwindsor.ca

"Eden" or, more commonly, the "Garden of Eden," made its first literal appearance in modern biblical texts. It was a place where everything grew, and all lived in harmony. It was irrigated by river water and mist since there was no rain. Initially, there were two unclothed ageless human custodians. But none of the texts mentioned waste. Perhaps there wasn't any? If so, could then this idyllic ancient land be replicated globally, for billions of human inhabitants, simply by reducing or eliminating the waste associated with anthropogenic activity and behavior, thus enabling them to become immortal in naturist nirvana? But where was it? Eden's most undisputed location is an area of modern-day

[1]This chapter is partly based on an unpublished keynote presentation given at the 2023 Reclaiming Eden – Engineering a Brighter Tomorrow Symposium, 22–23 June, University of Windsor, Ontario, Canada.

Reclaiming Eden: Responsible Living, Engineering, and Architectures
Edited by David S.-K. Ting and Jacqueline A. Stagner

ISBN 978-981-5129-31-1 (Hardcover), 978-1-003-49612-0 (eBook)
www.jennystanford.com

Iraq, making it hard to imagine 8–10 billion people living in just 0.34% of global land space. More realistically, could "Reclaiming Eden" be envisaged as a manifestation of the United Nations' vision of *Transforming our World* by 2030 into global habitats with *"Peace, Prosperity and Partnership beyond security"* for all? If so, will reducing waste, in the short time available, propel us toward this sustainability vision? It seems unlikely, yet by the end of the century, waste reduction will need to play a significant role in global efforts to provide more sustainable and prosperous communities. However, an accompanying vision is the mitigation of anthropogenic climate change by a complete global transition to renewable energy sources. This could lead to more waste being generated because of the disposal challenges associated with solar panels, electrochemical energy storage devices, and debris from the necessary increases in mining non-renewable metallic commodities. In terms of waste reduction, the two visions appear to be at odds. Thus, if the adequate and affordable provision of food, safe drinking water, and clean energy are to be realized, the visions will need to be reconciled, but how? A comprehensive written answer would need an encyclopedic approach so in this chapter the focus is limited to issues associated with municipal solid and food waste.

2.1 Introductory Remarks

Eden is mentioned, or alluded to, in many historical texts and is usually associated with the "Golden Age" of a particular civilization or nation. In all such cases, there is no mention of waste of any kind although there must have been some from all the fauna that lived in these regions. The geographical description of Eden given in the Holy Bible has led many, if not all, historians to identify its location somewhere in modern-day Iraq [1]. Most certainly, the texts placed the idyllic paradise between the rivers Tigris and Euphrates in an area of the ancient Fertile Crescent known as Mesopotamia (Fig. 2.1) [2]. Of the many descriptions of Eden and interpretations of the ancient texts perhaps the most explanatory is that provided in a late 19th century dictionary comparing Eden with the Greek's Golden Age [3], i.e., *life free from care, and without labour and sorrow. Old age was unknown; the body never lost its vigour; existence was a perpetual feast without a taint of evil. The earth brought forth spontaneously all things that were good in profuse abundance.*

An artist's impression is given in Fig. 2.2, where two custodians, Adam and Eve, of the Garden of Eden are shown naked, as stated in the biblical texts [4]. The inhabitants and the lush vegetation

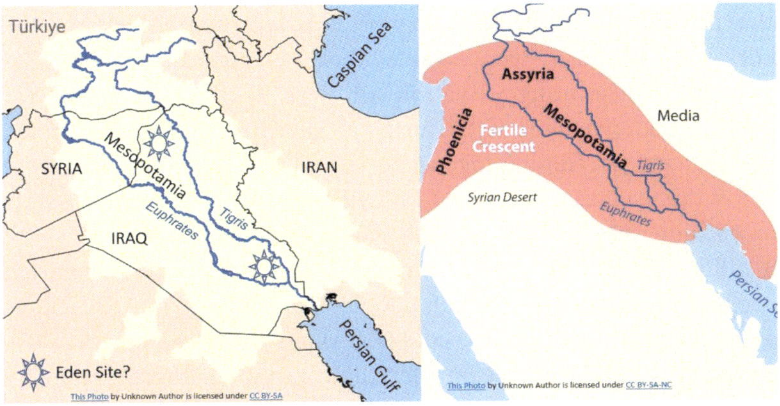

Figure 2.1 Eden and the Fertile Crescent.

Figure 2.2 Artist Impression of Eden: Thomas Cole [4].

flourished without any rain, obtaining water from the rivers and mist. This utopian environment came to an end when Adam was persuaded to eat the fruit from the tree of knowledge of good and evil, against the strict instructions of the garden's creator.

If the Garden was in Mesopotamia, then we know that it had an enviable climate and resources, until a few millennia ago, but then a changing climate caused the region to become hot and arid. This resulted in, perhaps, the first known instance of climate migration, as the people moved south toward the Persian Gulf. Is it conceivable that accumulating waste led to this deteriorating situation or was it a punishment for acquiring knowledge? If the original garden could be reclaimed by reducing or eliminating waste would the inhabitants have to detox themselves of acquired knowledge? But, if it were possible to reclaim Eden, could all the current global population, over 8 billion people, fit into the region? The necessary population density of just over 18,400 people/km^2, would be 200 times more than present-day Iraq and 300 times the global average [5]. However, while these seem incredible numbers there are many major cities whose population densities are greater, such as lower Manhattan, New York (Fig. 2.3) [6]. To achieve their intense densities, such cities have a large number of skyscrapers[2], making them seem more like concrete jungles than lush habitats, as shown in Fig. 2.2. It should also be acknowledged that in the original Eden, presumably, only one language, if at all, was spoken. Is it reasonable to assume that in a modern Eden, there would be a single lingua franca for communication?

Given the available data on the current success of climate mitigation and adaptation, and the efforts to produce affordable housing and electricity for all, it is perhaps self-evident that reversing millennia of changing climates and squeezing the global population into a sprawling and sizeable concrete habitat are seemingly impractical if not impossible tasks. However, if the mantra "Reclaiming Eden" is taken as a figurative concept or objective for transforming our planet for future generations then it is a worthy endeavor. To some extent, a strategic plan for achieving this transformation has already been developed as

[2]Defined as buildings over 40 stories high or 150 m.

described in the globally agreed United Nations (UN) 2015 resolution *"Transforming Our World: The 2030 Agenda for Sustainable Development"* [7], although it is noteworthy that it is now about four decades ago that a UN-commissioned report, "Report of the World Commission on Environment and Development: Our Common Future" highlighted the concept of sustainable development [8]. Since that time, there have been numerous attempts to define exactly what is meant by sustainable or sustainability, representing a range of versions but all have a common theme in harmony with the "Transforming the World" agenda, a typical example being that promoted by the USEPA, namely [9]:

> *"Sustainability – the continued protection of human health and the environment while enhancing economic prosperity and social well-being."*

Figure 2.3 Skyscrapers of Lower Manhattan [6].

The UN resolution specifies a plan of action to achieve 5Ps, i.e., People, Planet, Prosperity, Peace, and Partnership, by defining 17 sustainable development goals (SDG) associated with 169 targets and 231 unique measures or indicators of success. The word "waste" or similar, e.g., "wasteful," however, only appears in seven of the multitude of SDG targets one being, a somewhat verbose, international target calling for a reduction in the wasteful consumption of fossil fuels. But arguably almost all forms of waste generation characterize an inefficient use of materials from food products and drinking water to the detritus from building demolitions. Does then the paucity of focused waste reduction targets mean waste is not a key element in the transformation to a global sustainable future? Not at all, the lack of the frequency of use of the word "waste" largely belies the thrust of the UN's sustainable agenda target 12, "*Responsible Consumption and Production*," which embodies the organization's zero-waste global challenge. Humans and human activity produce a multitude of wastes. The form that is giving rise to most international concern is so-called municipal solid waste (MSW).

MSW is a collective term for at least 14 different waste streams including food, metals, textiles, and so on but MSW is not the only category of waste resulting from human activities. In fact, there are many such wastes but no universally applied classifications and definitions. Many nationally defined categorizations are encyclopedic in nature and therefore not readily distinguishable to the layperson. A relatively simple and useful waste category guide is perhaps used and publicized by the Environmental Protection Agency of the United States (USEPA), which categorizes waste into 11 groups, as shown in Table 2.1 [10]. In discussing how to reduce generated waste or, preferably, preventing its generation these categorizations can be helpful in providing the context. For example, the attention given to MSW can be appreciated by the fact that overall, the world generates 2.01 billion tons annually, which is set to rise to 3.4 billion tons by 2050, and its management alone could cost at least US$300 billion. Even this is not the worst-case scenario since the World Bank projections indicate the "Peak Trash" generation will only occur after the end of this century when the annual level will eventually reach over 4 billion tons per year [11]. While not all MSW ends up in landfill sites, the scale of the problem is illustrated in Fig. 2.4, which shows the height of natural and human-made structures compared with that of the depth of one year's worth of MSW deposited in a landfill [12].

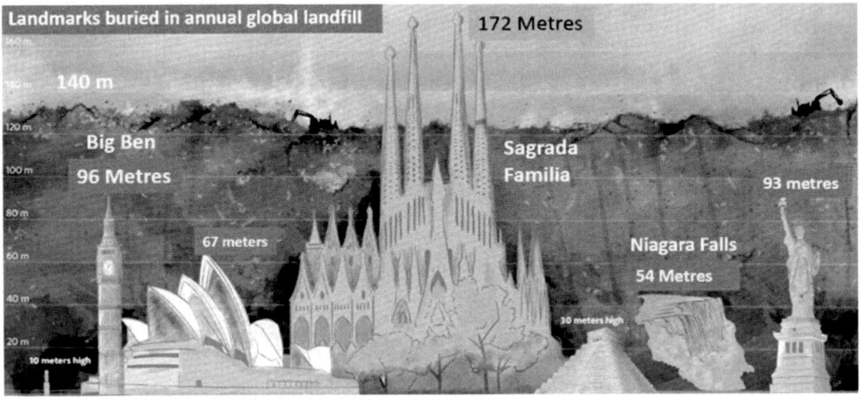

Figure 2.4 Illustration of annual MSW global landfill height with landmarks [12].

Table 2.1 USEPA waste categories [10]

USEPA waste categories	
MSW	Radioactive
Hazardous	Construction and demolition debris
Industrial non-hazardous	Extraction and mining
Agricultural and animal	Oil and gas production
Medical	Fossil fuel combustion
Sewage sludge	

Similar comparisons could be made with other types of waste, but in a short chapter, such as this, the main thrust of the waste-not-want-not theme focuses on MSW issues and particularly, but not exclusively, the food aspect of the UN's so-called "Water, Food and Energy" nexus [13]. The centuries-old word "nexus," originally used to describe a connection has taken on a more modern, perhaps somewhat overused, inference of direct linkage. The production of food and the generation of energy both require large quantities of water, usually freshwater if not drinking water. Moreover, global food production and distribution's annual energy needs are almost a third of the world's total energy consumption [14]. The main point of this chapter is that reducing food waste can also result in water and energy being saved. Yet only one of the SDG 12 targets has a physical number attached to it namely food waste. The target's aim is to reduce food waste at the retail and consumer level by 50% by 2030 [15]. Is food waste such a big issue? Yes. In the United States, for instance, over one-fifth of the material generated, which forms MSW and contributes to the need for landfill sites, is food waste.

Food is necessary for survival and while global production is more than sufficient to feed all almost 2 billion people, currently 1-in-4, are experiencing food insecurity and over 660 million are undernourished [16]. However, as post-pandemic food cost inflation has taken hold in many countries, both rich and poor, the 2019 data quoted in the previous sentence is likely to be an underestimate for 2023. Drinking water is also essential for survival and good health, more so than food, yet on the same scale as food insecurity, 1-in-4 do not have access to safe drinking water [17, 18]. Almost all food production comes from agricultural activities from plant croplands and animal pastures, and both need water. For crops the water comes from irrigation and precipitation and, although the quality can be less than that for human consumption, contaminated water can spoil the produce. There are stricter regulations for pasture animals, especially dairy herds, and many veterinarians recommend human standard quality for household pets [19, 20]. The food industries also have strict national regulations, including water quality, which are closely monitored during the food processing and preparation phases of the supply chain [21]. Could reducing food waste alleviate these daunting situations of food and drinking water insecurity?

Moreover, as energy is also part of the UN nexus, it is worth noting that water is also required in energy production, but care needs to be taken when estimating water use in the energy sector as withdrawn and consumed water are vastly different quantities. In a thermal powerplant, large quantities of water must be withdrawn from a suitable water source, e.g., a lake or river, but only a relatively small amount is consumed in the production of steam the remainder being recycled in some manner [22]. Solar energy converters also consume measurable amounts of water in cooling and cleaning processes, but some of this will eventually be recycled as a contribution to groundwater resources [23]. While there are many aspects to the use of water in energy conversion and power generation, a comprehensive discussion of these is outside the scope of this chapter other than to offer a word of caution about some apparently misleading information being circulated about water use in renewable energy systems [24, 25]. Moreover, a factor that is now attracting the attention of politicians, academics, and researchers is that MSW, especially when disposed of in landfills, becomes a major contributor to global greenhouse gas (GHG) emissions, as it decomposes anaerobically. Thus, if mitigating anthropogenic climate change is a primary objective of sustainable development then there is an urgent need to reduce waste. There are therefore numerous reasons to address the global waste issues. The challenges that will be encountered will likely be exacerbated as the world's population increases, especially if income levels rise in less affluent regions.

2.2 MSW and Its Constituents

In attempting to reduce waste a core issue is that generated by households and other anthropogenic activities. As previously mentioned, it can be anticipated that the amount of waste generated will increase with population, albeit somewhat in proportion to changes in annual incomes [26]. Indeed, using a high-income country like the United States as a well-documented example of population impacts, as the population has increased by a factor of 1.82 over the period 1960–2018, MSW generation[3] has increased by a factor of 3.32 [27, 28]. In other words, in this instance, over 80% more waste per capita was being produced, at least pre-pandemic, than in 1960. The composition and relative proportions of the MSW in terms of material types have also changed in the same time period. These changes often provide new challenges for the general handling and disposal of waste. While Paper and Paperboard waste still constitute the largest portion of MSW in weight, categories such as plastic and "other," which includes

[3]It should be noted that changing measurement methodologies may impact the precise value of these factors.

electrolytes in batteries, fluff pulps[4], and feces and urine in disposable diapers, have experienced the largest proportional increases [28]. Food waste has also increased but at a slightly lower rate than the growth in population, yet in 2018, it was the second-largest contributor to MSW. Until recently, an often-ignored contributor to MSW generation, was textiles, clothing, and footwear, all of which have increased and today they represent a growing cause for concern especially in lower-income countries, which are the recipients of large quantities of discarded fast-fashion items from higher-income countries [29].

However, merely reducing waste generation is only part of the problem because waste must also be collected, sorted, treated, and combusted with energy recovery, and, if not recycled or can't be recycled, it has to be disposed of in some manner, usually in landfills. But along the lines of the appellation that "one person's junk is another's treasure" an item discarded by one person could still be used by someone else or "repurposed" [30]. An example of this somewhat complex system is, perhaps surprisingly, the trillion-dollar fast fashion clothing industry. Individuals, especially in richer[5] countries, buy 60% more clothes today than they did a decade or so ago and maybe as much as 300% more than two generations ago [29]. The clothing industry is an emblematic example of the *make-use-discard* human mentality of the 21st century. After the clothes are discarded, they may be recycled in some manner such as through a thrift or charity shop, burned, or sent to landfill sites. However, many of these sites are not in the countries where the clothing items are purchased but are shipped to the African and Indian continents. At their destinations, they may be reused, recycled, or repurposed, but the bulk of the many million tons of original discarded clothing will be dumped in landfill sites. According to the World Bank, 10% of annual global carbon emissions come from the fashion industry – more than aviation and marine shipping combined – and could surge to 50% by 2030 [31]. Moreover, synthetic materials such as polyester, which is a petroleum by-product, i.e., plastic, are used extensively in

[4]Fluff pulps are used as a raw material in the absorbing parts of personal care products.

[5]The distinction between countries and individuals and gross national products and per capita incomes is not as straightforward as it may seem, as discussed later.

fast fashion clothing when they are discarded and eventually dumped into the rivers and oceans, they annually contribute the equivalent of 50 billion plastic bottles of pollution [32].

In the USEPA system, the materials generated in 2018 in the municipal waste stream, prior to any type of treatment, can be summarized by weight and proportion into the 11 types, as shown in Table 2.2 [33].

Table 2.2 Total generated materials in MSW [33] (author modified)

Material*§	Millions of tons	% of total waste
Paper & paperboard	67.4	23.1
Food	63.1	21.6
Plastics	35.7	12.2
Yard trimmings	35.4	12.1
Various metals	25.6	8.8
Wood	18.1	6.2
Textiles	17.0	5.8
Glass	12.3	4.2
Rubber and leather**	9.2	3.1
Other***	4.6	1.5
Misc. inorganic	4.1	1.4
Total waste******	**292**	**100**

§Before recycling, compositing, combusted, other food pathway management or landfilled.

*Not including construction and demolition debris, industrial process waste, and certain other wastes.

**Includes automobile and truck tires, carpets, rugs, and other miscellaneous durables.

***Includes battery electrolytes, fluff pulp, disposable diaper feces, and urine.

****Data round-up.

Not all MSW will be shipped to landfill sites, as illustrated in Fig. 2.5, using rubber and leather waste generation as an example [33]. In this case, just slightly less than 55% will eventually be disposed of in a landfill but this still represents 5 million tons of waste. The proportion of generated rubber and leather waste that reaches a purposely created landfill site is about the same as the overall percentage of municipal wastes in the United States that are disposed of in landfills. Globally this is the largest amount of landfill waste other than for the Latin America and Caribbean region. However, in Sub-Saharan Africa and South Asia, the landfill waste percentages are much lower at 24% and 4%, respectively, and as a result, the majority of the generated solid waste in these regions is merely dumped (Fig. 2.6) [26]. While many high- and middle-income countries are making some effort to better handle waste, for them the challenges will be far less onerous than in the dumping-reliant regions, as they are forecasted to produce greater rates of solid waste by mid-century [26].

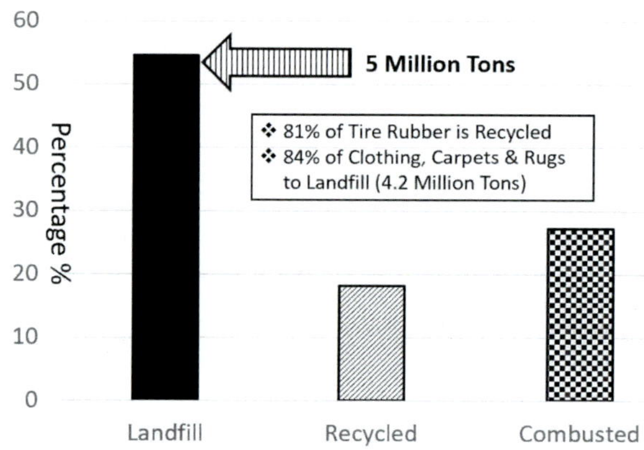

Figure 2.5 Management of rubber and leather generation in the United States [33].

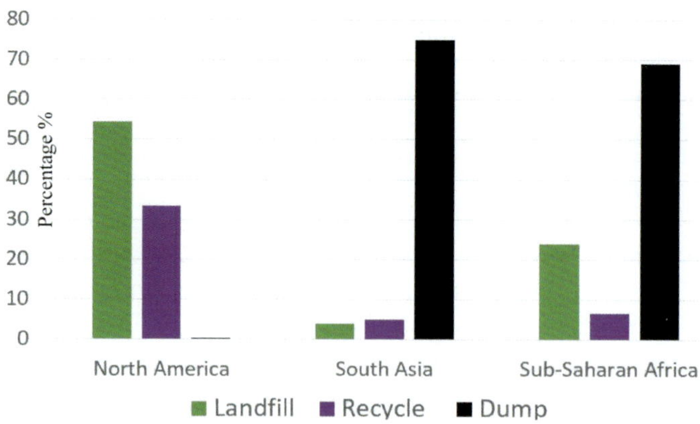

Figure 2.6 MSW management by region [26].

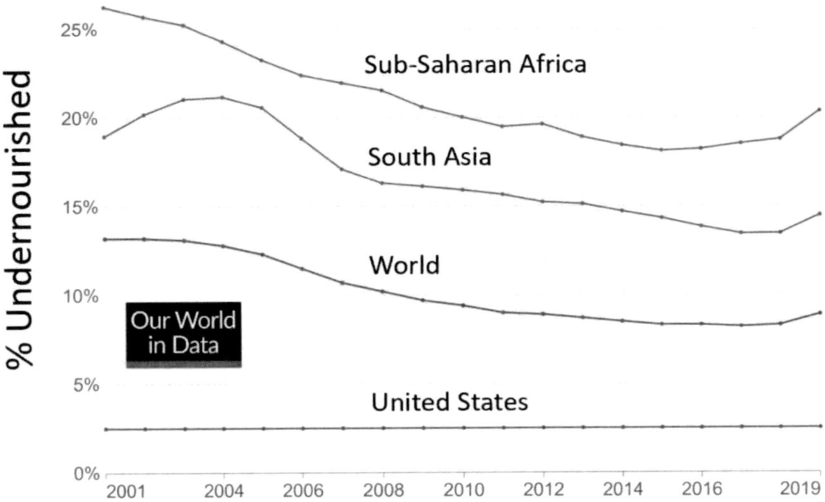

Figure 2.7 Exemplars of % undernourishment 2001–2019 [16].

At the same time, as Sub-Saharan Africa and South Asia are slated to produce more waste, the percentage of their populations suffering undernourishment, as defined by the UNFAO, is also set to increase (Fig. 2.7) [16]. The impacts of the COVID-19 pandemic and post-pandemic inflation on the data shown in the figure are yet to be determined but the relative effects of armed conflicts are demonstrated in the trajectories of undernourishment in Somalia and Yemen (Fig. 2.8) [16]. Reducing waste is unlikely to be instrumental in eliminating conflicts but a truly sustainable world will need to find alternatives to conflict resolution other than by war, something it has failed to do over the past 3 millennia in almost all cases. Notwithstanding the continuing need to take cognizance of the negative impacts of conflicts on global well-being, not resolving issues associated with imbalances in regional population growths will have an even greater impact on the waste factors involved in achieving sustainable development goals. For instance, the forecast population growth in Sub-Saharan Africa could mean that its total population, approximately the same as the combined total in Europe and North America in 2022, will be almost 90% higher by 2050 than the combined example [34]. This means that 1 in 5 people living on the Earth will reside in Sub-Saharan

Africa by 2050 and yet it is a region of high undernourishment and waste dumping. Can waste reduction, especially food waste, in say North America, which will represent 4% of the global population by mid-century alleviate the severe problems likely to be encountered by the 22% living in Sub-Saharan Africa? Will eliminating food waste in North America be sufficient to feed those in undernourished countries?

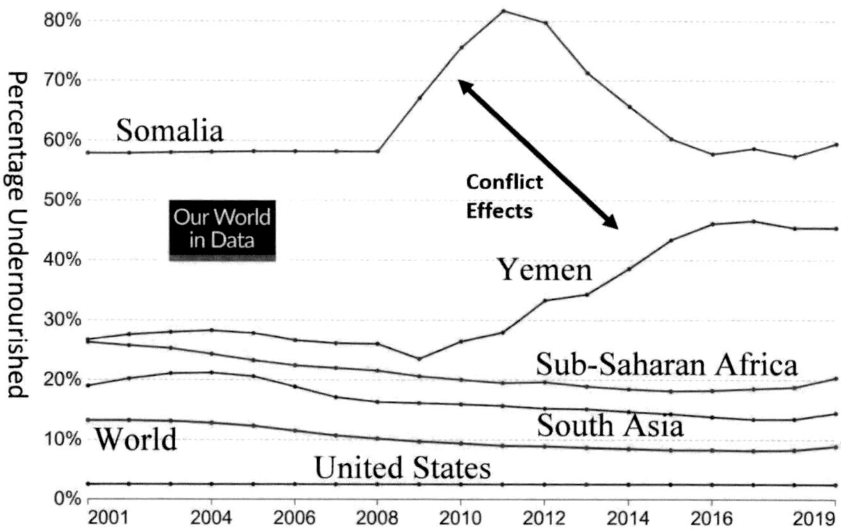

Figure 2.8 Percentage undernourished population including conflict zone exemplars [16].

2.3 Food Loss and Waste (FLW)

It was stated in the introductory remarks that in the United States over 20% of MSW is food waste. So, there are several other sources that contribute to the overall generation of MSW. Furthermore, not all food waste comes directly from residential sources. In some countries, an almost equal amount comes from commercial outlets such as hotels, restaurants, "fast-food" and fresh-food outlets, and supermarkets [10]. Even so, in the journey from production to consumption almost as much food can be wasted before it arrives at its final destination due to a multiplicity of reasons mainly associated with adverse weather and climatic effects as well as the vagaries of armed conflicts, inadequate land management, absence of infrastructure, and lack of electricity for refrigeration. To differentiate food wasted once it arrives at the consumer from that wasted before it does so, the commonly used term is "food loss" [35]. The United Nation's Food and Agriculture Organization (UNFAO) has formalized definitions for food loss and waste, which they refer to as FLW, namely [36]:

> *"Food loss: the decrease in the quantity or quality of food resulting from decisions and actions by food supply chain actors from the production stage up to, but excluding, retailers, food service providers, and consumers."*
>
> *"Food waste: the decrease in the quantity or quality of food resulting from decisions and actions by retailers, food service providers, and consumers."*

To determine a global portrait of the individual scales of loss and waste the UN, through the SDG 12 indicators, has introduced the "Food Loss Index" and the "Food Waste Index." The calculation of both requires member nations to produce the same data using universally standard methodologies but to date, such an implementation has proved problematic. Perhaps, this is not surprising since they are relatively new measures and depend upon a country's ability to collect and analyze accurate statistical data. Consequently, the central collection agency, the UNFAO, labels each country's input into classifications with very low confidence to high confidence, but only when national data actually exists, which is not always the case [37]. In the previous section, the question was posed about feeding the undernourished by reducing the waste generated by high-income countries, embodying an arguable presumption that hungry populations do not waste as much food as others, if at all. However, as shown in Table 2.3, comparing some of the UNFAO's data on annual food waste for individual countries, the answer is not straightforward.

Low-income countries like India, Ethiopia, and Nigeria produce far more food waste than high-income countries like the United States, the United Kingdom, and Saudi Arabia. Maybe this is because of the differences in population size? Not surprisingly, high-income Austria produces far less food waste than high-income Australia, which has a population threefold greater. But per capita, Nigeria generates three times the amount of food waste as the United States with a comparable population. Given that Nigeria is a low-income country, according to the World Bank, while the United States is a high-income country, one might expect that the waste generation rates would be reversed. There appear to be many contradictions in attempting comparisons between national food waste generation rates, population sizes, and incomes. These deserve intensive and extensive investigations, both national and international, if far greater insights are to be realized into the actions that need to be taken in reducing global food waste and seeking a sustainable future.

Table 2.3 Exemplars of UN Food Waste data and World Bank incomes

Country/region	Annual household food waste kg per capita*	Total households thousands of tons per year	Gross national income per capita**
Austria	39	349	High
South Africa	40	2,329	Medium
India	50	68,760	Low
United States	59	19,359	High
China	64	91,646	Medium
United Kingdom	77	5,200	High
Ethiopia	92	10,327	Low
Australia	102	2,563	High
Saudi Arabia	105	3,594	High
Malta	129	57	High
Greece	142	1,484	High
Nigeria	189	37,941	Low
Sub-Saharan Africa	40–189 (Average 108)		

*National Data with medium or high confidence.

**World Bank definitions based on their "Atlas" methodology.

The Food Loss Index (FLI), which is used as one of the indicators for SDG 12.3, is a measure of the losses from post-harvest on the farms to the start of the retail part of the food supply chain. As with the Food Waste Index (FWI), the methodologies used to determine the index, which is based on a value of 100 for 2015 for each country or region, are still a work in progress and, as the calculations are not without some complexities, the UN does offer training courses for the data gatherers [15, 38, 39]. Although the intention is to determine global indexes, each country determines its FLI, by identifying a "basket" of its 10 main food production commodities. The UN's SDG tracking "custodians" use the collected data to estimate regional FLIs expressed either using the 2015 base values or as a percentage comparing the physical losses with the amount produced (Fig. 2.9).

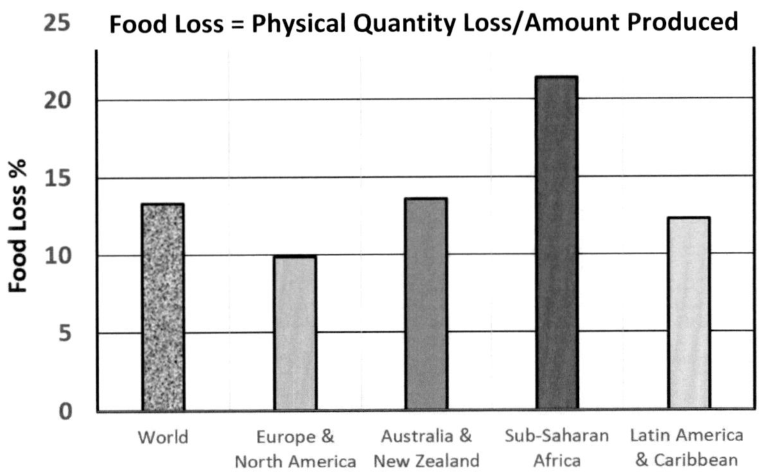

Figure 2.9 Exemplars of food loss index (%) using data from reference [39].

Arguably, the FLW measurement systems are still in their infancy, albeit improvements are being made continuously, and the global tracking of the involved indexes is providing useful information. For instance, once again as illustrated in Figs. 2.7 and 2.9, and Table 2.3, the region suffering the highest rate of undernourishment is also where the percentage of food loss is the greatest. Perhaps, if food waste and food loss in such a region could be reduced, then undernourishment would also be abated if not wholly eliminated. The challenge for the richer countries is how best to support such strategies in lower-income regions.

Equally, challenges regarding food production and consumption in richer countries cannot be ignored. National food and hunger indexes and the like provide an overview but to combat the specific occurrences of food waste and loss more information on the multitude of sources is necessary. Various government agencies and associate institutions in the United States, for example, while utilizing the UN's FLW protocols, have also developed additional investigative procedures to help clarify the choke points in the food production and consumption pathways. In the overarching labeling of food waste, they also identify specific entities such as "inedible" food which, although produced, is not intended for human consumption, and "excess" food which refers to food that is purposely donated to feed other people, such as, via a food bank. The amount of food donated in the United States in 2018 was over 7 million tons, enough to feed approximately 150,000 people for a week or to feed all of Canada's nightly homeless for over a month. Even so, in the same year, over 400,000 tons of food was discarded by food banks, as it became spoilt, or other difficulties were encountered, especially with distribution [40, 41]. The in-depth investigations and analyses of food waste in the United States are enabling detailed "Sankey" type plots of food waste generation and management pathways to be made, as shown in Fig. 2.10 [40].

As food is lost or wasted so too are other resources such as water, energy, and land used in food production as well as the labor employed in the food supply chain from farming to disposal. Indeed, food-water-energy-land-labor could be considered as an expansion of the UN's nexus, mentioned earlier, all of which are key elements in achieving a sustainable future. Moreover, as one of the major objectives of sustainability is improved human health, the link between food waste and the health risks of obesity needs to be brought to the attention of a wider audience. In a recent national health survey conducted in the United States, it was found that the prevalence of obesity existed in over 41% of the population with greater than 9% being severely obese [42].

The health costs are not only in terms of increased mortality rates but also financially for those suffering from obesity as treatment charges were almost US$173 billion more than those with a "healthy" body mass.

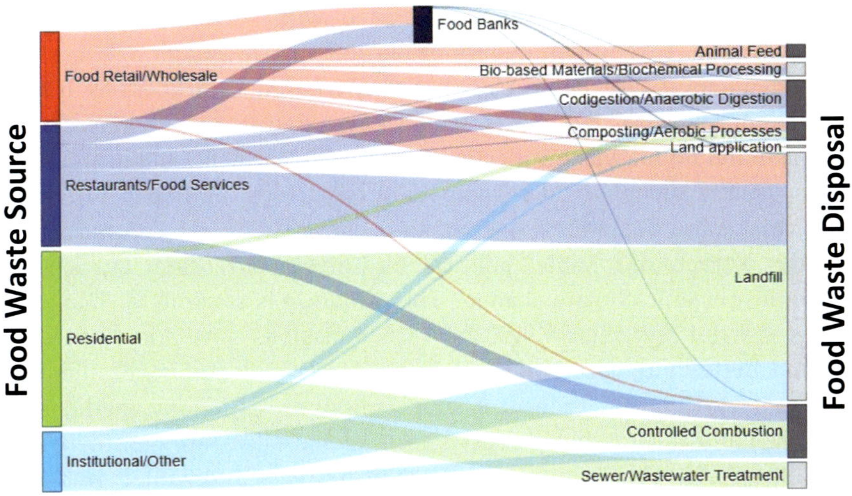

Figure 2.10 Sankey type plot of food generation and waste management [40].

In this chapter, FLW was used as an exemplar of global waste issues as food is required for survival, and therefore, for some, the lack of food now, or for many in the future, is worthy of awareness. The amount of food lost or wasted globally each year is astonishing, yet the same is true for many other forms of waste generated by human activities. That being the case the mantra waste not: want not, needs to be a general and continuing commitment, especially as a transition to renewable energies, using such devices as solar panels and battery storage, will generate new types of waste in addition to those already contributing to MSW, hazardous waste, and mining debris. Determining the scale of the global waste problems and their detrimental effects on sustainability, as defined earlier in this paper, could be effective in considering methodologies aimed at fulfilling the waste mantra. Consequently, some of the approaches to reducing or eliminating waste are discussed in the next section.

2.4 Reducing Waste

2.4.1 Lack of Awareness/Appreciation

Research shows that there is a major need to boost general awareness of the problems associated with waste, particularly with respect to human quality of life [43–45]. Although networks and alliances are being established increased awareness is not yet apparent [46, 47]. Why? Maybe as residential and commercial solid and liquid waste collection, handling, treatment, and disposal is usually the physical responsibility of a local authority rather than a state or national agency, the role of waste reduction does not attract the same political or media attention as, say, anthropogenic climate change? This situation is starting to change as the link between MSW and GHG emissions gets documented and better publicized, at least in academia and non-government agencies [26, 48]. Indeed, if such emissions, especially methane (CH_4) and carbon dioxide (CO_2), from global MSW, are considered collectively, they would be the third-largest individual global emitter source (Fig. 2.11) [49, 50]. However, this fact has not attracted much awareness within society. Is this because, especially in so-called high-income countries, community interest in waste is confined to knowing on what day it will be collected?

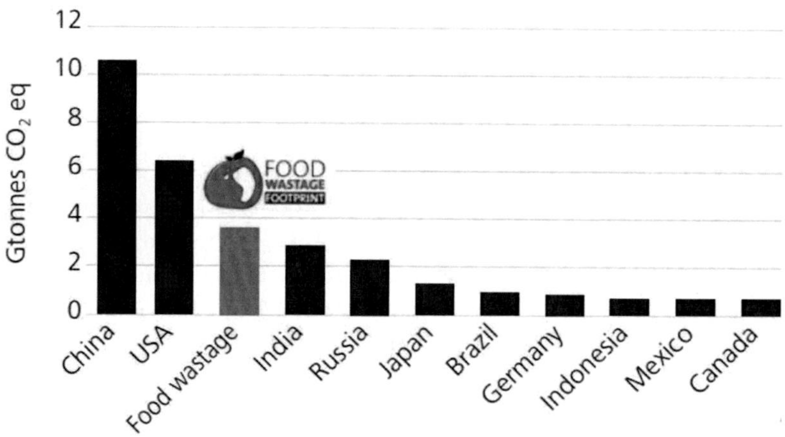

Figure 2.11 Comparison of global food waste GHG emissions [49, 50].

Only when a weather event or labor action prevents waste collection does the populace become interested and concerned, mainly because of the aesthetics of huge piles of garbage bags on the streets and their attraction to diseased vermin. Yet, despite an increase in voluntary recycling, albeit with local authority encouragement and taxpayer funding, the final destination of most MSW is still a landfill site. As such sites are usually "out-of-sight—out of mind" or, at least, hidden from residential communities, once again there is rarely a cause for public concern, but when it happens, all levels of government (national, state/provisional, and local) are involved [51, 52]. National or federal governments, and, sometimes, governmental conglomerations such as the European Union (EU), issue directives and regulations, state governments specify how the regulations are to be satisfied through the issuing of licenses, often to private companies, and municipal authorities ensure that the collection of waste and the associated landfill sites are operated within the regulatory framework [53].

The role of federal governments becomes more prominent when "hazardous" waste is involved and transported across jurisdictional boundaries, national, and international. Hazardous waste is any waste, not necessarily MSW, which has properties that are or could be dangerous or harmful to the environment and public health as defined by national environmental agencies or by the UN's Basel Convention when such waste is moved across regional or international borders [54, 55]. But in many incidents of waste mismanagement or regulatory breaches exactly who acts to remedy the problem and the timing of the action can be problematic involving a tortuous litany of legal proceedings [56–58]. The situation can be intractable if the problems are associated with landfills that are no longer in use and the contractors to whom the operations were outsourced no longer exist. The same type of circumstance occurs when forms of effluent contaminants leach into the ground from former industrial sites. Site remediation may be necessary but often legal proceedings on cause and effect, e.g., groundwater contamination, can be lengthy and difficult. Moreover, in almost all cases, it is the question of who pays that becomes contentious. Impacts of waste mismanagement are regular at the local level, meaning that the awareness of such problems is confined to those directly affected rather than the wider populace.

2.4.2 Sustainability Education

The generation of waste and what happens to it after it has been discarded is clearly a multifaceted subject. The extensive literature dealing with it is equally multi-layered ranging from government, industrial, and international reports to articles in specialized academic books, journals, and conferences as well as trade magazines and all the associated websites. What becomes apparent, when attempting a review of such published material, is that most is aimed at a specifically targeted audience, usually involved in some form of waste management and financing, and not engineers outside the civil and environmental disciplines although there is a wealth of literature from which general engineering courses and modules could be developed. The awareness of students in mechanical, industrial, system, electrical, and similar undergraduate programs, for example, the need to improve the design of equipment employed in waste management and its processing, and be cognizance of the logistics of the waste "supply-chain" could be of great benefit not only to their careers, to their instructors, but also society at large [59–61].

It could be anticipated that the inclusion of the engineering aspects of sustainability and sustainable development are now becoming core essentials of engineering curricula. But this is not quite the case although some acknowledgment of the concepts is slowly creeping into the accreditation criteria of engineering authorities such as the Engineering Accreditation Commission (EAC) of ABET[6] in the United States, which emphasize the inclusion of the concept of sustainability as part of the curriculum of an accredited engineering program dealing with engineering design [62, 63]. In Canada, "*Sustainable Development and Environmental Stewardship*" is one of seven topics that must be included in the complementary studies portion of the curriculum, which in total, constitute about 12% of the length of an accredited program, albeit sustainable development is mentioned

[6]Formerly the Accreditation Board for Engineering and Technology.

in terms of expecting that *"suitable committee and reporting structures are in place to assure the sustainable development and measurement of graduate attributes"* and that there is, *"a comprehensive, sustainable assessment plan for all attributes"* [64]. In the post-COVID era, it may well be that more focus will be placed on the engineering adaption of sustainability, as defined at the start of this chapter, and sustainable development goals in the UN's "Transforming Our World Agenda" [7].

From an engineer's viewpoint, the inclusion of sustainability education is not just desirable, but surely it should be part of the national commitments to the UN's transformation agenda SDG target 4, and specifically to SDG indicator 4.7, i.e., *"By 2030, ensure that all learners acquire the knowledge and skills to promote sustainable development,"* [65]. The framework of this relatively recent initiative would have, or should have, been well-known to UN member states as the "pathways to sustainability" were mentioned in Agenda 21, the official document of the 1992 Earth Summit in Rio de Janeiro [66]. After "Rio," just over a decade later in 2005 the UN's "Decade of Education for Sustainable Development" was announced and publicized, making available a useful framework for all educational providers and authorities [67]. It appears that the responses, where there were responses, came from educational groups mainly involved in K-12 learning such as the US Partnership for Education for Sustainable Development (USPESD) [68]. This partnership has developed a detailed curriculum map including learning outcomes and standards for pre-secondary education students and teachers [69]. The higher education sector was somewhat slower to respond in such detail to the curriculum needs of their learning offerings, but this is now changing, albeit the COVID-19 pandemic delayed some of the plans for the inclusion of sustainable development education [70, 71].

The progress being made in sustainable development education is monitored by UNESCO as the designated custodian and the organization has also been very active in disseminating guidelines for the learning and teaching aspects of the SDGs [72]. Overall, it is clear that awareness of the need for sustainable development can be well served by educational involvement from the K-12 level and beyond. Precisely how these ventures will help the reduction of wastes of all descriptions is not immediately obvious, for to date the progress made in achieving SDG indicator 4.7 has not been wholly encouraging, caused to some extent by the search for acceptable methodologies for the gathering, collation and analyzing of the appropriate data. Engineering awareness alone will not be sufficient to enable anthropogenic waste generation to be reduced but it has to be realized that industrial waste is generated in much higher quantities than MSW, especially in high-income countries. So, while increasing general awareness of waste generation and its role in sustainable development through education is highly desirable, engineering awareness must not be neglected, as highlighted in the recent "Berlin Declaration" [73].

2.4.3 Taxing Waste

If approaches, such as a wider foundation of education and more vigorous public information campaigns, do not yield the desired results then somewhat punitive tactics like a waste tax on garbage collection could be imposed, as these have proved successful. The Pay-As-You-Throw (PAYT)[7] schemes for which people are charged for the amount of "trash"[8] they "throw" out, using only authorized collection bags, are used by several thousand cities in the United States (Fig. 2.12) and have proved successful in reducing waste by as much as 30% in some communities [74, 75]. But could increasing burdens on taxpayers lead to more illegal dumping, "fly-tipping." Apparently, the lack of legal disposal options is more of a significant factor than cost, a situation prevalent at the local level. However, providing sufficient options will increase the financial expenditures of municipal governments, possibly causing a reduction in other services or once again an increasing tax burden on households, businesses, and industry. The PAYT may be a successful approach when the population can afford it but in areas where living affordability is questionable, the system would hardly be welcomed.

---

[7]Also known as Save Money and Reduce Trash (SMART).

[8]Waste containing no yard trimmings or recyclables or hazardous materials or construction debris.

Figure 2.12 Example of a PAYT trash bag [74, 75].

2.4.4 Culture and Behavior

As always, when discussing issues that are considered to be global it is easy to neglect that the world is made up of many different cultures. In some countries, China for example, it is considered impolite not to leave some food after a meal as this indicates that insufficient food was provided by the "host" to those consuming it [76]. Conversely in India and Japan, it is considered disrespectful to leave any food from a meal [77]. In many other countries, eating too much food necessary to provide sufficient energy for their daily human activity is a growing crisis. For example, in the United States, almost 2-in-5 people are considered to be obese based on their body mass index (BMI) [78, 79]. While the eating habits of various cultures may seem a somewhat trivial matter, the size of the respective populations of these country examples represent, between them, over 3.2 billion people or 40% of the total global population. So, although it could be said that some food is intentionally wasted in China, in the United States, it appears that many people consume too much food, and, even in India, regardless of the cultural norms there is a growing obesity problem with at least 70 million people suffering from morbid obesity [80, 81]. The overall result is that globally about the same number of people overindulge their food intake, or willfully waste some of their consumption, as experience food insecurity.

2.4.5 Global Adaptation of the Circular Economy

The reduction strategies so far outlined to some extent represent piecemeal approaches to be adopted at the national or state levels. A more fundamental approach proposed for adoption globally is a universal change in economic models from the present "linear economy" to a "circular economy" in which waste generation is not merely reduced but stopped. For engineers, long familiar with the philosophy of planned obsolescence, such a change would involve a "cradle-to-cradle" approach to design, manufacture, and operation [82]. According to the protagonists the key is to "decouple economic activity from the consumption of finite resources" [83]. Paraphrasing the Ellen MacArthur Foundation's strong advocacy for the approach: "There are three principles of the circular economy, eliminating waste, circulating waste, and regenerating nature with the overall approach being buttressed by a transition to renewable energy"[9] [83]. Even in diagrammatic form (Fig. 2.13), fully explaining the circular economy approach is not within the purview of this brief chapter [84]. Readers, however, may wish to familiarize themselves with concepts as they are attracting federal government-level attention, especially as a tool that could be useful in slowing anthropogenic climate change and the philosophy and practice of engineering design [85, 86]. The Federal Government of the United States, in particular, is developing a national strategy to bring about a "transformative change" to a circular economy [87]. A prime element in the implementation of this change is a "National Recycling Strategy" [88]. Recycling is one of the "R's" of sustainability.

---

[9]Defined in this case as "energy derived from resources that are not depleted on timescales relevant to the economy [83]."

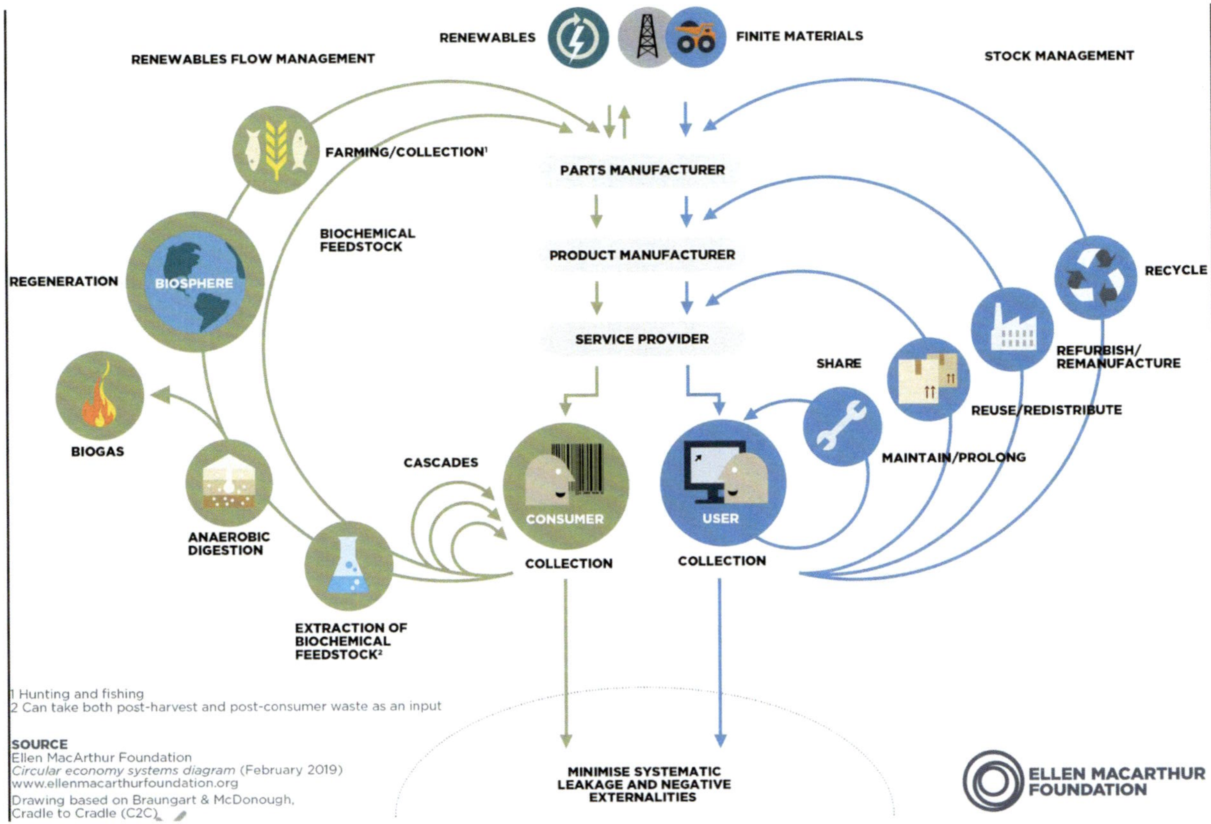

Figure 2.13 The Circular Economy MacArthur "Butterfly" diagram [84].

2.4.6 The "R's" of Sustainability

Prior to the 1970s, when the "3Rs" were mentioned, it was usually associated with educational matters, "Reading, Writing, and Arithmetic," but environmentalists redefined the 3R mantra to stand for "Reduce, Reuse, and Recycle" and it has now become a customary environmental term, especially when celebrating "Earth Day" [89]. By 2005, the 3Rs tag had been embraced by the United Nations Environmental Programme (UNEP) at the same time suggesting a 4th "R," repair, could also be added to a "life-cycle economy" approach [90]. This 4th "R" is now commonplace in sustainability literature although sometimes expressed as repair or repurpose, and further "R's" have been added to many lists such as rethink, refuse, revaluation, redistribution, relations, resilience, reassessment, restructuring, and rot (compost). When used in various combinations, this can yield a list as long as 13Rs, the most popular being 6Rs achieved by adding repair, rethink, and refuse, to the original 3Rs [91]. The 6R version is now introduced to students by grade school teachers up to tertiary education professors. It is a convenient and straightforward vehicle for communicating the key elements of the actions that need to be taken if global sustainability and the accompanying transformative change in the world economy are ever to be attained. In essence, it is a waypoint on the pathway to the adoption of a circular economy, but today there is a massive global "Circularity Gap" as less than 9% of the world has moved to the economic model, especially with regard to material usage [92]. Similarly, gaps in sustainability goal performance are also reported for reducing food waste and providing safe drinking water [10, 28, 37].

2.4.7 A Future "Hot-Spot" Sub-Saharan Africa

Having discussed problems and challenges of waste generation, particularly food waste, the implementation of many of the possible solutions, including awareness advocacy, pay-as-you-throw, enhanced sustainability development education, or the adoption of a circular economy, is unlikely to happen in any meaningful manner in regions such as sub-Saharan Africa in the foreseeable future. Why? As shown in Figs. 2.6, 2.7, and 2.9, this region dumps most of its waste, does little recycling, and has high rates of undernourishment and food loss. Many who live in the region are more likely to be concerned with where the next meal will come from rather than the need to tackle anthropogenic climate change. This is because of the 47/48 countries that constitute the region, over ¾ are considered to be least developed countries (LCD) or low-income food-deficit countries (LIFDC), or both, by the United Nations. In essence, this means these countries do not produce sufficient food for their populations, do not have the financial capacity to purchase imported food, have fragile infrastructure that impedes their ability for sustainable development, and are particularly vulnerable to the effects of conflict, inflationary food prices, damaging weather events, and societal inequities [93, 94]. Consequently, if tackling anthropogenic climate change and achieving the main aim of the agreed sustainable development agenda, i.e., the eradication of poverty, can only be accomplished by collective global effects, then those nations in a position to facilitate these outcomes need to help the nations who do not have the necessary capability. The challenges will be monumental.

If the forecasts are accurate, over the next three decades, the number of people living in sub-Saharan Africa is set to double representing 22% of the total global population, compared with just 4% for the United States, and with an accompanying three-fold increase in waste generation [26, 34]. Without collective global action, the efforts of the present high-income nations to mitigate climate change and accomplish sustainable development in their regions will likely prove to be futile globally and inadequate locally. Accordingly, how is this global action to be achieved and what are the best solutions for Africa? As can be anticipated, there are a variety of opinions ranging from increasing political and financial aid from foreign sources to the African States forming a European Union-type conglomerate [95, 96]. In 2015, the African EU concept was presented in detail in a document analogous to the UN's Agenda 2023, which the African Union labeled *"Agenda 2063 – Africa We Want,"* describing the publication as *"A shared strategic framework for inclusive growth and sustainable development & a global strategy to optimize the use of Africa's resources for the benefit of all Africans*" [96]. A key element of the framework is to create an African Free Trade Association, which is believed will provide the impetus for the industrialization of African economies with manufacturing playing a significant role. The problems of waste generation and management are also addressed in Agenda 2063 with proposed solutions following similar approaches to those of North America and Europe together with an ambitious target of 100% waste recycling by 2063.

A recent review of the possible approaches to reducing food waste in sub-Saharan Africa suggests that concentrating on traditional strategies to reduce post-harvest food losses may not only be prohibitively costly but also not the most cost-effective way of addressing the poverty and hunger issues prevailing in the region and offer four alternative objectives including to, "*increase profits along the food supply chain,*" [97]. Notwithstanding such discussions, the authors of the review highlight the fact that thorough analyses are not yet possible because of the meager amount of data available. In other publications, some aimed expressly at LCDs and LIFDCs, hunger and its relationships to conflict, poverty, national wealth, anthropogenic climate change, etc., are discussed with a view to developing methods to break these links but food waste is not dealt with explicitly or specifically mentioned [98, 99]. Consideration of the multitude of problems faced by sub-Saharan Africa is then not a neglected subject regionally or internationally, but actions would speak louder than words.

2.5 Concluding Remarks

It is only in the last decade or so that waste has become a higher priority on national and international governmental agendas. To gauge the scale of the potential waste problems significant effort has been directed toward the classification of the multitude of different waste streams and the establishment of universally accepted methodologies to measure the amounts generated by human activities. Despite the notable work being exercised in developing these methodologies, perhaps they can best be described as a work in progress. From the data and analyses that have been reported so far, it is clear that waste problems will continue to escalate in many low-income regions such as Sub-Saharan Africa. However, "Waste-Income-Population Size" correlations are complex and often contradictory, particularly with regard to food loss and food waste. Some regions with a high per capita rate of undernourishment also experience similarly high rates of food waste, whereas some with lower undernourishment and food waste rates are encountering increasingly high occurrences of obesity.

Many suggestions have been published, or publicized, on how waste could be reduced, from buying fewer clothes to adopting different diets, some more radical than others, or increasing taxes on waste generation and collection. It has emerged, perhaps surprisingly, that one of the main obstacles to achieving waste reduction targets appears to be more cultural than technical or economic. However, the curriculum of university-level engineering education with respect to sustainability and waste reduction could well benefit from a reappraisal. Given these circumstances, the complete elimination of waste by mid-century, proposed by the UN and as a founding principle of the adoption of a circular economy, seems overly optimistic. In the meantime, the best courses of action seem to be rapidly increasing global awareness of waste issues and seeking ways to overcome cultural inertia. The linkages between waste, GHG emissions, and anthropogenic climate change must surely give cause for concern if the sustainability agenda is ever to be achieved. Moreover, projections suggest that the world will not hit "Peak Trash" until the next century. So, while "Reclaiming Eden" may be a good pennant to unfurl in the cause of persuading humans to reduce wastage in support of achieving global sustainable development, any consequent actions are hardly likely to enable the "Reclaiming of Eden" by 2030 or even 2050, based on current performances.

Nevertheless, although Eden likely cannot be reclaimed, any efforts to reduce wastes, of all types, will not be a "waste of time.' Reducing waste will save land, energy, food, water, and money, and improve human health, as well as assist the mitigation and adaptation approaches involved in combating anthropogenic climate change. Despite the many suggestions for reducing or eliminating waste, arguably, of all the waste management strategies, perhaps the best one is for humans to learn not to generate it in the first place. Regrettably, even with the wholesale adoption of a circular economy, the establishment of a zero-waste world seems an unrealistic consequence.

In this chapter, the reasons for the need to reduce MSW and especially "Food Waste and Food Loss" have been the main focus, so it perhaps fitting to end this chapter with a quote from a wasted food "guru" Jonathan Bloom:

> *"Food waste isn't considered a problem because, for the most part, it isn't considered at all. It's easy to ignore because it's both common and customary. I have yet to meet someone who is pro-food waste, but many aren't convinced that it's important."* [100]

References

1. Bible Study Tools, "Eden," [Online]. Available: https://www.biblestudytools.com/dictionary/eden/. [Accessed 16 April 2023].
2. D. J. Hamblin, "Has the Garden of Eden Been Located at Last?" *Smithsonian Magazine*, Vol. 18 (2), May 1987. [Online]. Available: http://www.ldolphin.org/eden/. [Accessed 16 April 2023].
3. M. Easton, *Illustrated Bible Dictionary*, 3rd ed., Thomas Nelson, 1897.
4. T. Cole, Artist, *The Garden of Eden.* [Art]. Amon Carter Museum of American Art, Fort Worth, Texas, 1828.
5. Wikipedia, "List of Countries and Dependencies by Population Density," 23 March 2023. [Online]. Available: https://en.wikipedia.org/wiki/List_of_countries_and_dependencies_by_population_density. [Accessed 17 April 2023].
6. A. Quintano, "New York City," 2015. [Online]. Available: https://www.skyscrapercenter.com/city/new-york-city. [Accessed 17 April 2023].
7. United Nations, "Transforming Our World: The 2030 Agenda for Sustainable Development," Washington D.C., 2015.
8. G. H. Brundtland, *Our Common Future: The World Commission on Environment and Development*, 1st ed., Oxford University Press, 1987, p. 400.
9. USEPA, "ROE Glossary," 8 July 2022. [Online]. Available: https://www.epa.gov/report-environment/roe-glossary. [Accessed 17 April 2023].

10. USEPA, "Wastes: What Are the Trends in Wastes and Their Effects on Human Health and the Environment?" 2 March 2023. [Online]. Available: https://www.epa.gov/report-environment/wastes#roe-indicators. [Accessed 17 April 2023].
11. J. Stromberg, "When Will We Hit Peak Garbage," 30 October 2013. [Online]. Available: https://www.smithsonianmag.com/science-nature/when-will-we-hit-peak-garbage-7074398/. [Accessed 15 January 2023].
12. Sustainability for Oil: Acciona, "Landfills: A Serious Problem for the Environment," 2019. [Online]. Available: https://www.activesustainability.com/environment/landfills-serious-problem-environment/?_adin=11551547647. [Accessed 17 April 2023].
13. United Nations, "The Water-Energy-Food Nexus," Rome, 2014.
14. United Nations, ""Energy-Smart" Food for People and Climate," United Nations Food and Agriculture Organization, Rome, 2011.
15. One Planet Network, "SDG 12 Hub: Target 12.3 Food Loss & Waste," 2023. [Online]. Available: https://sdg12hub.org/sdg-12-hub/see-progress-on-sdg-12-by-target/123-food-loss-waste. [Accessed 17 April 2023].
16. M. Roser and H. Ritchie, ""Hunger and Undernourishment," Our World in Data," 2019. [Online]. Available: https://ourworldindata.org/hunger-and-undernourishment. [Accessed 17 April 2023].
17. H. Ritchie and M. Roser, "Clean Water," Our World in Data, 2021.
18. World Health Organization (WHO) and the United Nations Children's Fund (UNICEF), "Progress on household drinking water, sanitation and hygiene 2000–2020: five years into the SDGs.," Geneva, 2021.
19. USEPA, "Water Quality Standards Handbook," Washington, D.C., 2022.
20. A. Tague, "Cat & Dog Drinking Water: Which Water Sources Are Safe for Pets?" 27 May 2022. [Online]. Available: https://www.hillspet.com/pet-care/nutrition-feeding/pet-safe-water-sources. [Accessed 18 April 2023].
21. FAO and WHO, "Safety and Quality of Water Used in Food Production and Processing – Meeting report," UNFAO, Rome, 2019.
22. M. A. Harris and T. H. Dieh, "Withdrawal and Consumption of Water by Thermoelectric Power Plants in the United States, 2015," U.S. Department of the Interior, U.S. Geological Survey, Reston, Virginia, 2019.
23. S. V. K. K. Panat, "Electrostatic Dust Removal Using Adsorbed Moisture–Assisted Charge Induction for Sustainable Operation of Solar Panels," *Science Advances,* vol. 8, no. 10, p. 11, 11 March 2022.

24. C. Clarke, "Fact Check: How Much Water Does Solar Power Really Use?" 6 May 2014. [Online]. Available: https://www.kcet.org/redefine/fact-check-how-much-water-does-solar-power-really-use. [Accessed April 17 2023].

25. International Energy Agency (IEA), "Energy and Water: Exploring the Interdependence of Two Critical Resources," 2023. [Online]. Available: https://www.iea.org/topics/energy-and-water. [Accessed 18 April 2023].

26. S. Kaza, L. C. Yao, P. Bhada-Tata and F. Van Woerden, "What a Waste 2.0: A Global Snapshot of Solid Waste Management to 2050, Urban Development," World Bank, Washington, D.C., 2018.

27. World Bank Group, "Population, Total," 2023. [Online]. Available: https://data.worldbank.org/indicator/SP.POP.TOTL. [Accessed 18 April 2023].

28. USEPA, "National Overview: Facts and Figures on Materials, Wastes and Recycling," 2023. [Online]. Available: https://www.epa.gov/facts-and-figures-about-materials-waste-and-recycling/national-overview-facts-and-figures-materials#Trends1960–Today. [Accessed 18 April 2023].

29. M. Bedat, Unravelled, The Life and Death of a Garment, New York, New York: Portfolio/Penguin, 2021, p. 312.

30. K. Wells, "Purging Old Clothes? Here's How to Responsibly Donate and Repurpose Your Castoffs," 18 July 2022. [Online]. Available: https://www.nytimes.com/wirecutter/guides/how-to-get-rid-of-old-clothes/. [Accessed 18 April 2023].

31. World Bank Group, "How Much Do Our Wardrobes Cost to the Environment?" 23 September 2019. [Online]. Available: https://www.worldbank.org/en/news/feature/2019/09/23/costo-moda-medio-ambiente. [Accessed 19 April 2023].

32. E. Galley, "Fashion's Impact on our Oceans," 8 June 2020. [Online]. Available: https://www.fashionandthefree.com/fashion-impact-oceans. [Accessed 19 April 2023].

33. USEPA, "Advancing Sustainable Materials Management: 2018 Fact Sheet," Washington D.C., 2020.

34. United Nations Department of Economic and Social Affairs, "World Population Prospects 2022: Summary of Results," United Nations, New York, 2022.

35. World Food Program USA, "Food Waste vs. Food Loss: Know the Difference and Help #StopTheWaste Today," 2023. [Online]. Available:

https://www.wfpusa.org/articles/food-loss-vs-food-waste-primer/. [Accessed 19 April 2023].

36. UNFAO, "Voluntary Code of Conduct for Food Loss and Waste Reduction," FAO, Rome, 2022.
37. United Nations Environment Programme, "Food Waste Index Report 2021," United Nations, Nairobi, 2021.
38. C. Fabi and A. English, "Methological Proposal for Monitoring SDG Target 12.3. The Global Food Loss Index design, data collection methods and challenges," UNFAO, Rome, 2018.
39. UNFAO, "SDG Sub-Indicator 12.3.1.a – Food Loss Index," 2020. [Online]. Available: https://elearning.fao.org/course/view.php?id=605. [Accessed 19 April 2023].
40. USEPA, "2018 Wasted Food Report: Estimates of Generation and Management of Wasted Food in the United States in 2018," USEPA, Washington, D.C., 2020.
41. Government of Canada, "Everyone Counts 2018 Highlights: Preliminary Results from the Second Nationally Coordinated Point-in-Time Count of Homelessness in Canadian Communities," Government of Canada, Ottawa, 2019.
42. Centers for Disease Control and Prevention (CDC), "Adult Obesity Facts," U.S. Department of Health & Human Services, 17 May 2022. [Online]. Available: https://www.cdc.gov/obesity/data/adult.html. [Accessed 19 April 2023].
43. Z. Han, Q. Duan, Y. Fei, D. S. G. Zeng and M. Hu, "Factors that Influence Public Awareness of Domestic Waste Characteristics and Management in Rural Areas," *Integrated Environmental Assessment and Management,* vol. 14, no. 3, pp. 395–406, 8 February 2018.
44. National Parks Conservation Association, "Study Reveals Lack of Awareness of Waste Challenges Facing US National Parks," 16 August 2016. [Online]. Available: https://www.npca.org/articles/1292-study-reveals-lack-of-awareness-of-waste-challenges-facing-us-national. [Accessed 19 April 2023].
45. S. E. Hasan, "Public Awareness Is Key to Successful Waste Management," *Journal of Environmental Science and Health, Part A,* vol. 39, no. 2, pp. 483–492, 2004.
46. K. Steinbiß, E. Fröhlich and J. Sander, "Managing Sustainable Consumption: Shaping the Customer Journey with a Focus on Sustainability in the Food Industry," *EDUCATION OF ECONOMISTS AND MANAGERS,* vol. 62, no. 4, pp. 79–100, 2021.

47. United Nations Department of Economic and Social Affairs (UNDESA), "Zero Waste Challenge. 12. Responsible Consumption and Production, #SDGAction42688," 2023. [Online]. Available: https://sdgs.un.org/partnerships/zero-waste-challenge-12-responsible-consumption-and-production. [Accessed 17 April 2023].

48. H. Ritchie, "How Much of Global Greenhouse Gas Emissions Come from Food?" Our World in Data, 18 March 2021. [Online]. Available: https://ourworldindata.org/greenhouse-gas-emissions-food. [Accessed 19 April 2023].

49. UNFAO, "Food Wastage Footprint & Climate Change," Rome, 2015.

50. M. Crippa, E. Solazzo, D. Guizzardi, F. Monforti-Ferrario and A. Leip, "Food Systems Are Responsible for a Third of Global Anthropogenic GHG Emissions.," *Nature Food,* vol. 2, pp. 198–209, March 2021.

51. Government of Canada, "Municipal Solid Waste: A Shared Responsibility," Government of Canada, 28 January 2022. [Online]. Available: https://www.canada.ca/en/environment-climate-change/services/managing-reducing-waste/municipal-solid/shared-responsibility.html. [Accessed 19 April 2023].

52. "https://en.wikipedia.org/wiki/Solid_waste_policy_of_the_United_States," 16 January 2023. [Online]. Available: https://en.wikipedia.org/wiki/Solid_waste_policy_of_the_United_States; Solid_waste_policy_of_the_United_States.pdf. [Accessed 19 April 2023].

53. C. Antico, "Brief Overview of the EU Waste Law and Policy," European Commission, Brussels, 2020.

54. UNEP, "Basel Convention on the Control of Transboundary Movements of Hazardous Wastes and their Disposal," Nairobi, 1989.

55. M. Putter, "What Is Hazardous Waste?" CPD Online College Limited, 16 December 2022. [Online]. Available: https://cpdonline.co.uk/knowledge-base/health-and-safety/what-is-hazardous-waste/. [Accessed 20 April 2023].

56. D. Kellenberg, "The Economics of the International Trade of Waste," *The Annual Review of Resource Economics,* vol. 7, pp. 109–125, 3 June 2015.

57. S. Colvin, "Whose Liability Is It Anyway? Residual Liability for Historic Landfill Sites," 2023. [Online]. Available: https://emlawshare.co.uk/resource/whose-liability-anyway-residual-liability-historic-landfill-sites/. [Accessed 20 April 2023].

58. O. Barsalou and M. H. Picard, "International Environmental Law in an Era of Globalized Waste," *Chinese Journal of International Law,* vol. 17, no. 3, pp. 887–906, September 2018.

59. J. K. Debrah, D. G. Vidal and M. A. P. Dinis, "Raising Awareness on Solid Waste Management through Formal Education for Sustainability: A Developing Countries Evidence Review," *Recycling 2021, 6(1),* vol. 6, no. 6, p. 21, 22 January 2021.
60. B. Przestrzelski, S. M. Lord and M. M. Camacho, "Trash Teachings: How a Materials Science Module Series about Waste can Empower Engineering Students to be More Sociotechnically Responsible," in *American Society for Engineering Education, 126th Annual Conference & Exposition, 2019,* Tampa, Florida, 2019.
61. The United Nations Educational, Scientific and Cultural Organization, (UNESCO), "Issues and trends in Education for Sustainable Development," Paris, 2018.
62. M. Abd-Elwahed and A. Al-Bahi, "Sustainability Awareness in Engineering Curriculum Through a Proposed Teaching and Assessment Framework," *International Journal of Technology and Design Education,* vol. 31, pp. 633–651, 2020.
63. ABET, "Criteria for Accrediting Engineering Programs 2023–2024," Baltimore, MD, 2022.
64. Engineers Canada, "2021 Accreditation Criteria and Procedures - Engineers Canada," Ottawa, 2021.
65. UNESCO, "Sustainable Development Goal 4 and its Targets," 2021. [Online]. Available: https://en.unesco.org/education2030-sdg4/targets. [Accessed 1 July 2023].
66. UN, "United Nations Conference on Environment & Development AGENDA 21," New York, 1992.
67. UNESCO, "UN Decade of Education for Sustainable Development, 2005–2014: the DESD at a glance," UN, New York, 2005.
68. USPESD, "https://uspartnership.org/pages/brief-history," 2023. [Online]. Available: https://uspartnership.org/pages/brief-history. [Accessed 1 July 2023].
69. USPESD, "US Partnership for Education for Sustainable Development National Education for Sustainability K-12 Student Learning Standards Version 3," 2009.
70. D. Galán-Casado, A. Moraleda, M. L. Martínez-Martí and M. Á. Pérez-Nieto, "Sustainable Environments in Education: Results on the Effects of the New Environments in Learning Processes of University Students.," *Sustainability,* vol. 12, no. 7, p. 12, 28 March 2020.
71. M. Fia, K. Ghasemzadeh and A. Paletta, "How Higher Education Institutions Walk Their Talk on the 2030 Agenda: A Systematic Literature Review," *Higher Education Policy,* p. 34, 24 June 2022.

72. UNESCO, "Education for Sustainable Development Goals: Learning Objectives," Paris, France, 2017.

73. UNESCO, "Berlin Declaration on Education for Sustainable Development; Learn for Our Planet: Act for Sustainability," Paris, 2022.

74. L. B. Pollans, "What Is Pay-As-You-Throw? A Waste Expert Explains," 4 January 2022. [Online]. Available: https://theconversation.com/what-is-pay-as-you-throw-a-waste-expert-explains-173828. [Accessed 20 April 2023].

75. State of Connecticut, "https://portal.ct.gov/DEEP/Reduce-Reuse-Recycle/Payt/Save-Money-and-Reduce-Trash," April 2021. [Online]. Available: https://portal.ct.gov/DEEP/Reduce-Reuse-Recycle/Payt/Save-Money-and-Reduce-Trash. [Accessed 20 April 2023].

76. F. Ingrossa, "Will Deep-Rooted Habits Affect China's "Clean Plate" Campaign?" DAO Insights, 14 October 2020. [Online]. Available: https://daoinsights.com/opinions/will-deep-rooted-habits-affect-chinas-clean-plate-campaign/. [Accessed 20 April 2023].

77. J. R. Thomson, "If You Don't Finish Your Plate in These Countries, You Might Offend Someone," The Huffington Post, 10 October 2016. [Online]. Available: https://www.huffpost.com/entry/empty-plate-cultural-differences_n_5807822de4b0dd54ce368d7e. [Accessed 20 April 2023].

78. National Public Health Information Coalition (NPHIC), "Obesity More Deadly than Previously Believed," NPHIC, 29 March 2023. [Online]. Available: https://www.nphic.org/news/featured-topics/1365-obesity-more-deadly-than-previously-believed. [Accessed 20 April 2023].

79. B. Stierman, J. Afful, M. Carroll, T.-C. Chen and O. e. a. Davy, "National Health and Nutrition Examination Survey 2017–March 2020 Pre-pandemic Data Files—Development of Files and Prevalence Estimates for Selected Health Outcomes," U.S. Department of Health and Human Services, 2021.

80. International Institute for Population Sciences (IIPS) and ICF, "National Family Health Survey (NFHS-5), 2019–21: India Volume 1," Mumbai, 2022.

81. Wikipedia contributors, ""Obesity in India," Wikipedia, The Free Encyclopedia," 25 January 2023. [Online]. Available: https://en.wikipedia.org/wiki/Obesity_in_India. [Accessed 21 April 2023].

82. P. Özkan and Y. E. Karataş, "Chapter 4 – Linear Economy to Circular Economy: Planned Obsolescence to Cradle-to-Cradle Product

Perspective," in *Handbook of Research on Entrepreneurship Development and Opportunities in Circular Economy*, Hershey/Beijing, IGI Global, 2020, pp. 61–86 (Chapter) Book 696.

83. Ellen MacArthur Foundation, "What Is a Circular Economy?" [Online]. Available: https://ellenmacarthurfoundation.org/topics/circular-economy-introduction/overview. [Accessed 21 April 2023].
84. Ellen MacArthur Foundation, "The Butterfly Diagram: Visualising the Circular Economy," February 2019. [Online]. Available: https://ellenmacarthurfoundation.org/circular-economy-diagram. [Accessed 21 April 2023].
85. USEPA, "What is a Circular Economy?" 29 September 2022. [Online]. Available: https://www.epa.gov/circulareconomy/what-circular-economy. [Accessed 21 April 2023].
86. E. Delaney, W. Liu, Z. Zhu, Y. Xu and J. S. Dai, "The Investigation of Environmental Sustainability Within Product Design: A Critical Review," *Design Science*, vol. 8, no. e15, p. 43, 6 April 2022.
87. USEPA, "Building a Circular Economy for All: Progress Toward Transformative Change," EPA, Washington, D.C., 2022.
88. USEPA, "National Recycling Strategy Part One of a Series on Building a Circular Economy for All," EPA, Washington, D.C, 2021.
89. T. Ayoob, "Reduce, Reuse & Recycle: Where Does the Phrase Come from and What Does It Actually Mean?" 16 June 2022. [Online]. Available: https://thelatch.com.au/history-of-recycling/. [Accessed 23 April 2023].
90. E. Director and UNEP, *Reduce, Reuse and Recycle Concept (the "3Rs") and Life-Cycle Economy*, Nairobi, UNEP, 2004.
91. Acciona Sustainability for all, "#LEARNSUSTAINABILITY: The 3RS, 6RS and 9RS," 2019. [Online]. Available: https://www.activesustainability.com/sustainable-life/learnsustainability-the-3rs-6rs-and-9rs. [Accessed 24 April 2023].
92. Circle Economy, "Global Circularity Gap Report 2023," Circle Economy/PACE, Amsterdam, 2023.
93. UN DESA, "LDC Identification Criteria & Indicators," [Online]. Available: https://www.un.org/development/desa/dpad/least-developed-country-category/ldc-criteria.html. [Accessed 1 July 2023].
94. UNFAO, "Low Income Food Deficit Countries (LIFDC) Query," United Nations, [Online]. Available: https://data.apps.fao.org/catalog/dataset/special-country-groups/resource/2967dcfe-aa26-44af-afde-e97ca5e30450. [Accessed 1 July 2023].

95. R. Munang and Z. Han, "Food security: Regional solutions key to solving Africa's challenges," [Online]. Available: https://www.un.org/africarenewal/web-features/food-security-regional-solutions-key-solving-africa%E2%80%99s-challenges. [Accessed 1 July 2023].

96. African Union, "Agenda 2063 The Africa We Want," African Union, Addis Ababa, 2015.

97. M. Sheahan and C. B. Barrett, "Review: Food loss and waste in Sub-Saharan Africa," *Food Policy,* vol. 70, p. 12, 24 April 2017.

98. UN DESA, "Handbook on the Least Developed Country Category: Inclusion, Graduation and Special Support Measures," United Nations, New York, 2021.

99. C. Delgado, K. Tschunkert and D. Smith, "Food Insecurity in Africa: Drivers and Solutions," Stockholm International Peace Research Institute, Sipri, Stockholm, 2023.

100. J. Bloom, "https://www.trvst.world/waste-recycling/food-waste/food-waste-quotes/," 25 June 2022. [Online]. Available: https://www.trvst.world/waste-recycling/food-waste/food-waste-quotes/. [Accessed 25 April 2023].

Chapter 3

Harbor-House Design with Recycled Insulation Materials for Sustainable Development

Figen Balo,[a] **Hazal Boydak,**[b] **and Lutfu S. Sua**[c]

[a]*Department of METE, Engineering Faculty, Firat University, Turkey*
[b]*Department of Architecture, Dicle University, Diyarbakır, Turkey*
[c]*Department of Management and Marketing,*
Southern University and A&M College, Louisiana, USA

figenbalo@gmail.com

The building information modeling (BIM) methodology used in the construction, architecture, and engineering sectors may be a vital way to optimize building efficiency and decrease the sectors' ecological effects. Given the request for more energy-efficient constructions, it is significant to use the BIM-based BEM method. Thus, designers are able to compare and identify diverse sustainable scenarios at the first step of the plan, thereby conserving money, time, and other resources. Due to the rising environmental concerns over the last several years, sustainable and green constructions have become more popular. Buildings now account for a sizeable portion of global energy use and are a substantial source of greenhouse gas emissions throughout

Reclaiming Eden: Responsible Living, Engineering, and Architectures
Edited by David S.-K. Ting and Jacqueline A. Stagner

ISBN 978-981-5129-31-1 (Hardcover), 978-1-003-49612-0 (eBook)
www.jennystanford.com

the course of their lifetimes. Constructing green buildings can be a laborious process to implement several sustainable design criteria, but incorporating BIM into the planning can make the process easier.

In this research, BIM is used to evaluate the sustainability of a harbor house's construction. In the city of Çanakkale, Turkey, this study is done to investigate the energy-efficiency benefits of using different building and insulation materials for constructing a green harbor home. In addition, this study focused on the contribution of environmental waste to energy efficiency in buildings by considering recycled insulation materials. According to the investigation, there was an energy savings of around 2.44% between the highest and the lowest energy-efficient types when compared to designed types with different insulation and construction materials. Therefore, sustainability judgments about the design can be made using this method since various design possibilities may easily be developed within the conceptual phase itself.

3.1 Introduction

Sustainable improvement's aim is described as an improvement that contains today's needs without endangering the next generations' capability to fulfill their requirements [1]. This aim was also set as precedence in the sector since the conventional method of building, planning, demolishing, and operating a building utilizes a lot of non-renewable resources and energy, and produces a lot of trash [2, 3]. Several criteria govern the development of a sustainable structure: strive for quality in the built environment, increase resource reuse, reduce resource consumption, use recyclable or renewable materials, protect the natural ecology, use recyclable or renewable materials, increase resource reuse, reduce resource usage, preserve the ecology, and establish a non-toxic and healthy environment [4]. BIM can apply and promote maintainable methods in building designs, monitor building efficiency, and analyze and collect data to help designers make decisions at various phases of a project [5, 6].

The complexity and difficulty of construction planning are increasing. The BIM has arisen to optimize, modernize, and automate the conventional operations of the architecture-engineering-construction business, and it is already affecting the existing building project situation. Several design teams in various nations are embracing this method because of its advantages over traditional design methodologies. In addition, several governments are advocating it as a standard design approach. In the United Kingdom, for example, BIM has been required in all public designs since the beginning of 2016 [7].

The BIM methodology may be characterized as a combination of processes, policies, and industries that enable the management of digital three-dimensional (3D) drawings and other design data throughout a construction's life cycle. It is an integrated technique for managing a plan's functional and physical qualities while numerically repeating the plan's real building [8]. This model shift has ramifications and advantages not just for the building industry, but generally for the community, like better constructions that use fewer materials, demand fewer people and financial resources, and run more efficiently [9]. BIM's core is the exchange of data among project stakeholders across the whole construction lifecycle. This permits each stakeholder to study independently while remaining integrated into a central model that houses overall construction-sourced data, such as engineering/technical and architectural designs. This helps planners to rapidly detect incompatibilities and flaws, allowing them to make better decisions while also saving money and time [10].

When this technology is used on a construction project, it produces a digital model that is comprised of all the plan's attributes. Object-oriented parametric modulation is used to create this model, which implies that characteristics are described to characterize the relationship between objects (for example columns, beams, pipes, ducts, and so on). A BIM platform – a simulation programmer that produces various utilizes data and develops and modifies BIM models – is suggested for creating the virtual model. BIM may be utilized over the course of the whole construction lifecycle, including documentation, visualization, management, maintenance, cost planning, comprehensive analysis, operation, destruction, logistics, and so on [11, 12].

Many of these applications require the usage of extra BIM tools, which are various types of software that do certain sorts of analysis. Information may be exchanged between the BIM and these tools in a variety of pathways. Different interchange file formats, like Green-Building-XML and sector Foundation-Classes, might be utilized depending on the study purpose. Direct interfaces to connect various BIM platforms and technologies have been developed.

Despite the many advantages of BIM, only 0.36 of building enterprises in Europe use it [13]. This may be due to the trouble of implementing novel methodologies inside organizations, which is commonly owing to resistance to shifting from old pathways and is exacerbated by a lack of comprehension of the topic [14]. Other probable causes include the software's professional certification and high cost. Interoperability issues across software, as well as a lack of standardization, impede BIM deployment [15–17]. However, BIM's use in sustainable building has increased exponentially, owing to its capability to simplify building efficiency analysis and sustainability [18]. Some BIM studies in recent years are given in Table 3.1.

Table 3.1 Some of the BIM-based studies in recent years

Research topic	Research method	Year	Source
BIM green perspective	Modeling	2015	[19]
Green BIM	Review	2015	[20]
Building orientating impact on the construction energy usage	Case study and modeling	2016	[21]
Carbon and BIM calculations h	Case study and modeling	2016	[22]
Combination of LCA and BIM	Case study and modeling	2017	[23]
Post-structure energy-efficiently analyzes	Review	2017	[24]
Combination of LCA and BIM	Case study and modeling	2017	[25]
Carbon computations and BIM	Modeling h	2017	[26]
GHG computations	Case study and modeling	2017	[27]
Holistically BIM perspective for lesser-carbon planning	Case study and modeling	2018	[28]

Research topic	Research method	Year	Source
Combination of LCA and BIM	Case study and modeling	2018	[29]
Data-based sustainable planning	Modeling	2019	[30]
BIM's interoperability	Review	2019	[31]
Combination of LCA and BIM	Case study and modeling	2019	[32]
Combination of LCA and BIM	Case study and modeling	2020	[33]
Combination of LCA and BIM	Case study and modeling	2020	[34]
Building external wall	Modeling	2020	[35]
Energy efficiency and maintainability	Case study and modeling	2020	[36]
Green building and BIM evaluation	Integrated research	2020	[37]
Carbon emissions and entire-life costs	Case study and modeling	2020	[38]
Combination of LCA and BIM	Review	2020	[39]
Smart buildings	Integrated research	2020	[40]
Combination of LCA and BIM	Case study and modeling	2020	[41]
Holistic green BIM	Review	2021	[42]
Renewable-material choice	Case study and modeling	2021	[43]
Mounted concrete constructions	Case study and modeling	2021	[44]
Building C-footprint prediction	Modeling	2021	[45]
Green building and BIM assessment	Interviews	2021	[46]

This study aims to use Autodesk Revit and Autodesk Green Building Studio to assess the design of a green harbor house in Çanakkale, Turkey, for energy efficiency analysis. The building energy performance analysis performed in this study focuses on energy demand according to heating and cooling loads. In addition, when recycled insulation materials are evaluated in terms of energy efficiency, their effects on performance are analyzed. The main contribution of this study is to evaluate the alternative building and recycled insulation materials used in exterior walls in terms of energy consumption using Green Building Studio simulation.

3.2 Methodology

The building sector is liable for a variety of negative environmental effects, resulting in an increase in requests for sustainable construction. There are a few construction maintainability evaluation methodologies available that enable planners to assess and optimize the grade of building sustainability. Because of the BIM's potential, construction sustainability assessment methodologies can benefit from BIM. XML connectors and Green Building Studio connect their analysis equipment to major BIM resolutions such as AutoCAD MEP, AutoCAD Architecture, and Revit Architecture. With this capacity, researchers can utilize the improved data generated by the BIM to test construction efficiency and validate planning alternatives more effectively on the Internet.

This research primarily makes use of the Autodesk programs Revit Architecture and Green Building Studio. High-quality, energy-efficient buildings may be designed, built, and maintained with the aid of Revit. The Revit model can quickly be changed to reflect any changes in the design, making the design process and documentation simpler and more dependable. Green Building Studio employs the DOE 2 simulation engine and is a versatile cloud-based solution. It is flexible and precise equipment that enables engineers, architects, and planners to enhance energy efficiency and work toward carbon neutrality.

The structure studied in this project is a harbor house designed in the Turkish city of Çanakkale. Green construction techniques and various building insulation materials are researched, and the necessary drawings and other data are gathered. Different rooms and places are assigned after the model is developed, along with an occupancy number. The spaces are then divided up into various zones. The building's areas, zones, and overall accuracy are verified in the "Heating and Cooling Load Settings".

To generate the energy modeling in Revit Architecture, the energy settings are defined. For creating the software modeling for the energy performance assessment in Revit Architecture, there are two options. The first method involves utilizing "notional masses", which allow the consumer to assess a design's notional step. The alternative is to embrace "building components."

Building components is the option chosen for the assessment because the design and other information are available. This model is then exported in gbXML format to "Green Building Studio" for additional assessments, such as an entire construction energy assessment, and for comparison with the energy consumption of a harbor house. Type files were created with certain properties such as materials and material thicknesses for the building model created according to the project designed by Autodesk Revit 2021. The files created for each of these types were transferred to the "Green Buildings Studio" program in gbXLM format. Then, energy analysis was made for each type and the results were evaluated.

A new project is created in the Green Building Studio online window by entering the project name, facility type, location, operating schedule, currency, time zone, and weather station. The data are then changed; otherwise, the evaluation will use default values. The gbXML file for the building is uploaded once the project has been built. After that, it is run, and the outcomes are shown in the Green Building Studio interface. The outcomes of the "Base Run" are those produced in accordance with the created types. Additionally, GBS creates additional "Design Alternatives" for the model.

3.2.1 Design Stages of Harbor House

In this study, a harbor house was modeled in Autodesk Revit 2021 considering the climatic conditions of Canakkale (Turkey) [Latitude: 40.166, Longitude: 26]. Annual design conditions according to the climate of Canakkale city are given in Table 3.2. The wind values of Canakkale city are displayed in Fig. 3.1.

Table 3.2 Annual design conditions according to the climate of Canakkale city

	Cooling degree day		Heating degree day	
	Threshold	**Value**	**Threshold**	**Value**
	18.3 °C	680	18.3 °C	1998
	21.1 °C	359	15.6 °C	1409
	23.9 °C	133	12.8 °C	905

(*Continued*)

Table 3.2 (*Continued*)

Threshold	Cooling		Heating	
	Dry bulb (°C)	MCWB (°C)	Dry bulb (°C)	MCWB (°C)
0.1%	36.0	22.0	-6.6	-7.0
0.2%	35.3	22.1	-6.2	-6.7
0.4%	34.3	21.9	-5.8	-6.4
0.5%	34.1	21.8	-5.8	-6.4
1%	33.1	21.0	-4.8	-5.5
2%	32.3	20.8	-3.1	-4.0
2.5%	31.8	20.8	-2.3	-3.6
5%	30.0	19.8	0.2	-0.9

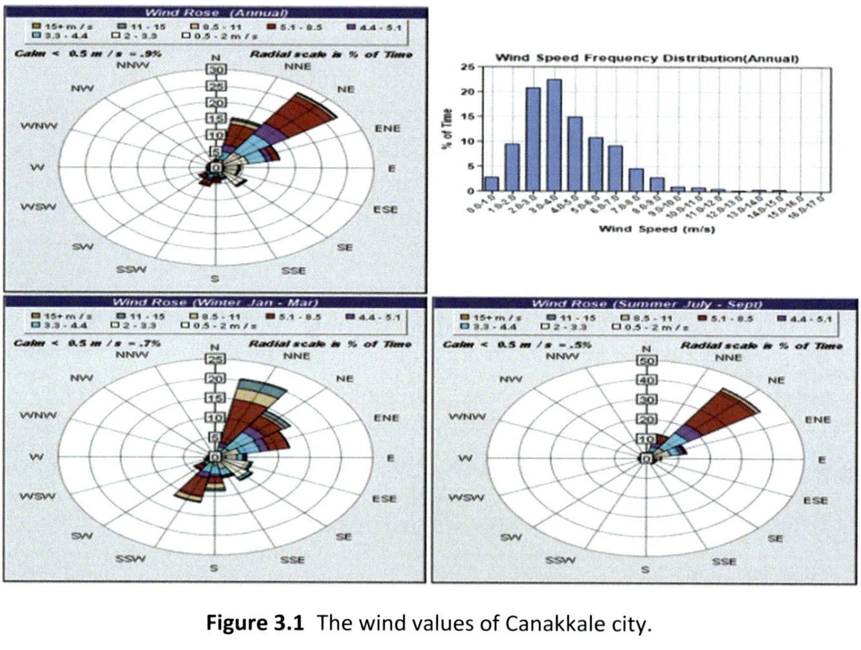

Figure 3.1 The wind values of Canakkale city.

In the design, three different building materials (pumice stone, pumice, and brick), four different recycled insulation materials (recycled glass fiber, recycled PET, recycled textile, and recycled cotton), and three different wall types (external insulation, internal insulation, and sandwich wall) were used.

Twelve different types of walls were created by using different construction and insulation materials for each wall type used. A total of 36 different alternatives for all wall types were realized in different combinations with Autodesk Revit simulation. Each type is designed for certain exterior wall layers (with roof, structure, and insulation materials) and certain wall structures (external insulation, internal insulation, or sandwich walls). For example, for Type-1 (internally insulated wall), it was modeled as "plaster + pumice stone + recycled glass fiber + plastic" by Autodesk Revit 2021, and then by the "Green Buildings Studio" program, energy analysis was performed with these materials, and a solution is obtained for Type-1 when the building roof was a "green roof". The green roof is used as the roof material in all designs. Externally insulated wall designs are Type-1 through Type-12, internally insulated wall designs are Type-13 through Type-24, and sandwich wall structure designs are Type-25 through Type-36. The amount of energy consumed by the harbor house in one year was analyzed by using building energy modeling in the Green Building Studio simulation for all types. The exterior and interior views obtained by Lumion software of the harbor-house design are displayed in Fig. 3.2. The building information of the harbor-house design is given in Table 3.3. The analyzed wall structures for the harbor-house design are shown in Fig. 3.3.

Table 3.3 Harbor-house building information

Carrier system	Masonry construction
Number of floors	Ground floor
Story height	3 m
Dimensions	10 m × 10 m
Gross area	100 m^2
Net area	87 m^2
Wall thickness	30 cm

The building's energy efficiency is impacted by material selection. By increasing the building's thermal insulation, the energy required to heat and cool a structure can be significantly lowered [48–50]. Geometrical parameters, material's thermal qualities, environmental circumstances, and building use are all important determinants of thermal performance [51, 52]. As a result, the project's materials were chosen based on their thermal characteristics and their environmental advantages. The optimum insulation thickness for Çanakkale province of Turkey was taken from the literature as 6 cm and used in the simulation [53]. Alternative types and the thermo-physical properties of building envelope components of the designed harbor house are given in Table 3.4 and Table 3.5, respectively. U values and thickness layers of building components are given in Table 3.6 and Table 3.7, respectively.

Figure 3.2 The exterior and interior views obtained by Lumion software of the harbor-house designed.

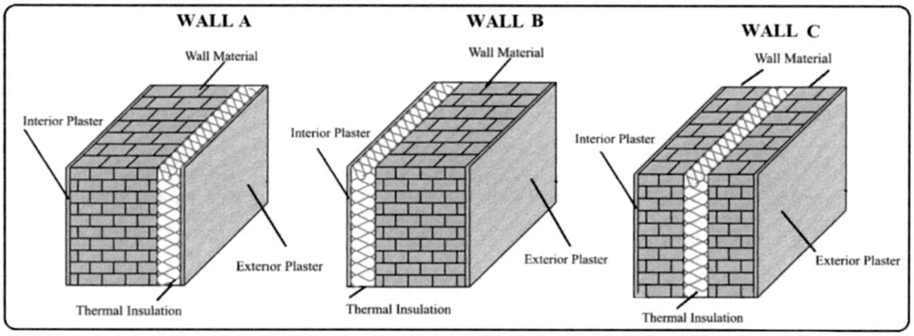

Figure 3.3 The analyzed wall structures for the harbor-house design.

Table 3.4 Alternative types of designed harbor houses

Type	Wall (cm), (In-Out)		
Type-1	Plaster (2.00)	Pumice stone (20.00) Recycled glass fiber (6.00)	Plaster (2.00)
Type-2	Plaster (2.00)	Bims block (20.00) Recycled glass fiber (6.00)	Plaster (2.00)
Type-3	Plaster (2.00)	Brick (20.00) Recycled glass fiber (6.00)	Plaster (2.00)
Type-4	Plaster (2.00)	Pumice stone (20.00) Recycled PET (6.00)	Plaster (2.00)
Type-5	Plaster (2.00)	Bims block (20.00) Recycled PET (6.00)	Plaster (2.00)
Type-6	Plaster (2.00)	Brick (20.00) Recycled PET (6.00)	Plaster (2.00)
Type-7	Plaster (2.00)	Pumice stone (20.00) Recycled textile (6.00)	Plaster (2.00)
Type-8	Plaster (2.00)	Bims block (20.00) Recycled textile (6.00)	Plaster (2.00)
Type-9	Plaster (2.00)	Brick (20.00) Recycled textile (6.00)	Plaster (2.00)
Type-10	Plaster (2.00)	Pumice stone (20.00) Recycled cotton (6.00)	Plaster (2.00)
Type-11	Plaster (2.00)	Bims block (20.00) Recycled cotton (6.00)	Plaster (2.00)
Type-12	Plaster (2.00)	Brick (20.00) Recycled cotton (6.00)	Plaster (2.00)
Type-13	Plaster (2.00)	Recycled glass fiber (6.00) Pumice stone (20.00)	Plaster (2.00)
Type-14	Plaster (2.00)	Recycled glass fiber (6.00) Bims block (20.00)	Plaster (2.00)
Type-15	Plaster (2.00)	Recycled glass fiber (6.00) Brick (20.00)	Plaster (2.00)

Table 3.4 (*Continued*)

Type	Wall (cm), (In-Out)		
Type-16	Plaster (2.00)	Recycled PET (6.00) Pumice stone (20.00)	Plaster (2.00)
Type-17	Plaster (2.00)	Recycled PET (6.00) Bims block (20.00)	Plaster (2.00)
Type-18	Plaster (2.00)	Recycled PET (6.00) Brick (20.00)	Plaster (2.00)
Type-19	Plaster (2.00)	Recycled textile (6.00) Pumice stone (20.00)	Plaster (2.00)
Type-20	Plaster (2.00)	Recycled textile (6.00) Bims block (20.00)	Plaster (2.00)
Type-21	Plaster (2.00)	Recycled textile (6.00) Brick (20.00)	Plaster (2.00)
Type-22	Plaster (2.00)	Recycled cotton (6.00) Pumice stone (20.00)	Plaster (2.00)
Type-23	Plaster (2.00)	Recycled cotton (6.00) Bims block (20.00)	Plaster (2.00)
Type-24	Plaster (2.00)	Recycled cotton (6.00) Brick (20.00)	Plaster (2.00)
Type-25	Plaster (2.00)	Pumice stone (6.00) Recycled glass Pumice stone (10.00) fiber (10.00)	Plaster (2.00)
Type-26	Plaster (2.00)	Bims block Recycled glass Bims block (10.00) fiber (6.00) (10.00)	Plaster (2.00)
Type-27	Plaster (2.00)	Recycled glass Brick (10.00) Brick (10.00) fiber (6.00)	Plaster (2.00)
Type-28	Plaster (2.00)	Pumice stone (10.00) Recycled PET (6.00) Pumice stone (10.00)	
Type-29	Plaster (2.00)	Bims block (10.00) Recycled PET (6.00) Bims block (10.00)	
Type-30	Plaster (2.00)	Brick (10.00) Recycled PET (6.00) Brick (10.00)	
Type-31	Plaster (2.00)	Pumice stone (10.00) Recycled textile (6.00) Pumice stone (10.00)	
Type-32	Plaster (2.00)	Bims block (10.00) Recycled textile (6.00) Bims block (10.00)	
Type-33	Plaster (2.00)	Brick (10.00) Recycled textile (6.00) Brick (10.00)	
Type-34	Plaster (2.00)	Pumice stone (10.00) Recycled cotton (6.00) Pumice stone (10.00)	
Type-35	Plaster (2.00)	Bims block (10.00) Recycled cotton (6.00) Bims block (10.00)	
Type-36	Plaster (2.00)	Brick (10.00) Recycled cotton (6.00) Brick (10.00)	

Table 3.5 Thermophysical properties of building envelope components

	Thermal conductivity [W/(m K)]	Density [kg/m^3]
Wall materials		
Pumice stone	0.18	250
Bims block	0.23	770
Brick	0.5	1200
Plaster	1.0	1800
Insulation materials		
Recycled glass fiber	0.031	450
Recycled PET	0.034	60
Recycled textile	0.040	50
Recycled cotton	0.039	25
Slab materials		
Slab on grade–Reinforced concrete slab	2.5	2400
Bedding mortar–Reinforced screed	1.4	2000
Protective concrete	1.65	2200
XPS	0.04	35.00

Table 3.6 U values of building components

Floor slab	U value: 0.7729 W/(m^2 K)
Suspended slab	U value: 8.6420 W/(m^2 K)
Green roof	U value: 0.1205 W/(m^2 K)

Table 3.7 Thicknesses and layers of building components

Floor slab (cm)	
Slab on grade (50.00) Waterproofing Bedding (0.10) mortar (5.00) XPS (4.00)	Protective concrete (3.00) Reinforced screed (5.00) Granite covering (1.20)
Suspended slab (cm)	
Wall paint (0.10) Plaster (2.00) Reinforced concrete slab (15.00)	Reinforced screed (5.00)
Green roof (cm)	
Vapor Control Layer (0.125 mm) Mineral Titanium Protan SE Wool (5.00)	Extensive Green Roof Layers (sedum/drainage) (11.5)

3.3 Results and Discussion

The energy performance analysis by BIM-based Green Building Studio simulation is conducted on the model to determine the most energy-efficient building components. It is important to combine the materials used in the layers that make up the building envelope in a way that contributes the most to energy efficiency. Among the types created by arranging the building and insulation materials layer by layer according to the wall structure, we call the type that gives the lowest output in terms of energy use as the optimum type. While the right decisions provide the maximum comfort conditions for the users, they also result in the efficient use of energy and resources. Different material characteristics were researched and analyzed concerning the building's energy efficiency in Çanakkale, Turkey. Generally, different material characteristics analyzed for the design display show that the energy efficiency of the harbor-house modeling has created a green construction. Green buildings (also known as sustainable buildings) mean that a building is compatible with the environment throughout its life cycle (planning, design, construction, operation, maintenance, renovation, and demolition) and uses natural or waste-converted resources efficiently. In this study, green and sustainable materials were used in terms of materials by using green roofs as the roofing material and recycled insulation materials as the insulation material. In addition, the type that consumes less energy was aimed to be determined. This way, less emission to the environment will be achieved by using less energy. Depending on the conclusions acquired, it seems that the energy usage is changed in the harbor-house modeling according to the building and insulation material types. The energy consumption values obtained by Green Buildings Studio software of the harbor-house design are given in Fig. 3.4.

Considering the annual cooling loads, the highest energy consumption was found to be 1747 kWh for Type-21 and Type-24. The lowest energy consumption was achieved for Type-2 and Type-26 at 1722 kWh. Considering the annual heating loads, the highest energy consumption was obtained for Type-21 at 9179 kWh. The lowest energy consumption was found for Type-25 to be 8154 kWh. When the annual total energy consumption is evaluated, the maximum energy consumption is 43090 kWh for Type-21 and the minimum energy consumption is 42038 kWh for Type-25.

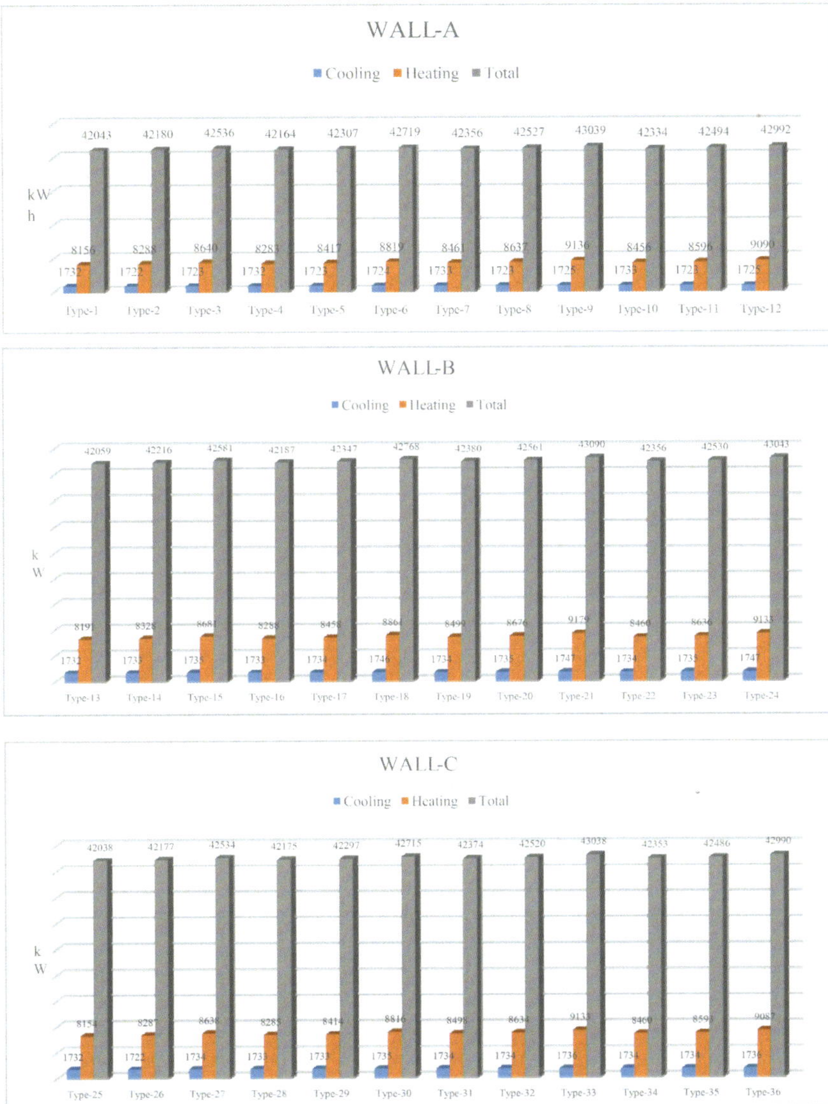

Figure 3.4 The energy consumption values obtained by Green Buildings Studio software of the harbor-house designed (Wall A: external insulated wall, Wall B: internal insulated wall, Wall C: sandwich wall).

Considering the annual energy consumption for cooling, annual heating energy consumption, and annual total energy consumption, the highest energy consumption was for the internally insulated wall design.

The annual maximum energy consumption is achieved by using internally insulated wall structures, brick as wall material, and recycled textiles as insulation material.

The annual minimum energy consumption was when the sandwich wall structure and pumice stone were used as the wall material and recycled glass fiber was used as the insulation material. Figure 3.5 shows how much energy is consumed for each "Type" of a wall, owing to the utilization of diverse building and insulation materials.

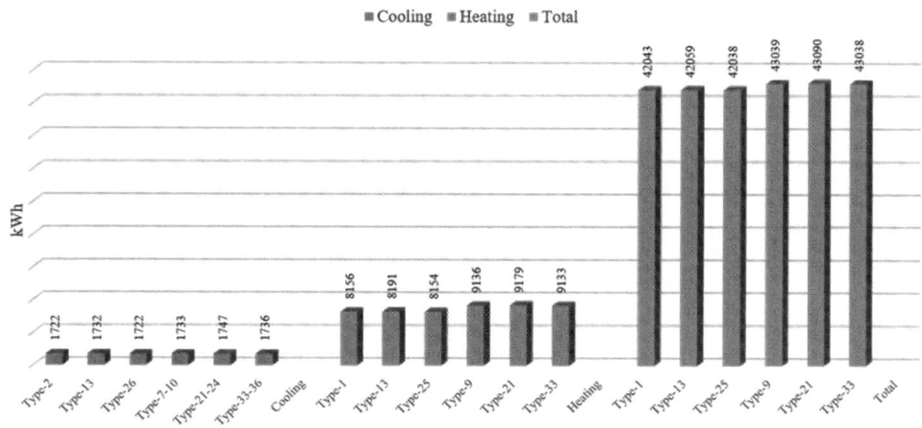

Figure 3.5 Annual energy consumption amounts for the most positive and negative alternatives of all scenarios.

In Fig. 3.6, annual energy consumption amounts for the most positive and negative alternatives of all scenarios are given. The first figure shows the electric end-use for cooling while the second one shows the use for heating.

Energy end-use values of Type-21 as the worst alternative according to most energy usage are displayed in Fig. 3.6. In Type-21, the misc. equipment and pump-aux consume the maximum and minimum amounts of energy, respectively.

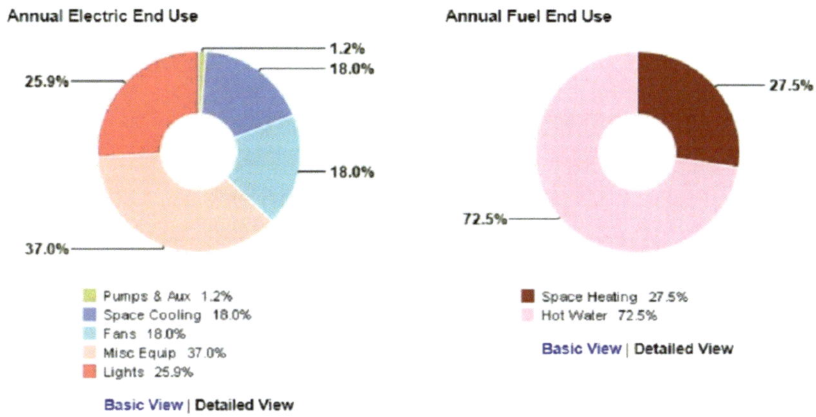

Figure 3.6 Energy end use of Type-21.

The annual energy cost values of the harbor-house design scenarios are illustrated in Fig. 3.7. The "Green Buildings Studio" program calculates the costs automatically per kilowatt hour. One kilowatt hour means one kilowatt of power (kW) consumed per hour and simply indicates how much energy an electrical appliance consumes. Here, it refers to the electricity consumption cost required for the 1 kWh of energy consumed by the building. The maximum annual energy cost, USD 158.81, belongs to Type-10 (recycled cotton as insulation material and pumice stone as building material, in an externally insulated wall structure). The reason for the high-cost calculations for Type-10 is due to the difference between electricity and fuel costs. Since the electricity cost is higher than the fuel cost, Type-10 came to the fore with a minimal difference in the results. The minimum annual energy cost, USD 155.84, is obtained for Type-25 (recycled glass fiber as insulation material and pumice stone as building material, in a sandwich wall structure).

Life-cycle cost values of the designed harbor house are shown in Fig. 3.8. The maximum life-cycle cost, USD 2162.62, is for Type-21 (recycled textile as insulation material and brick as a building material, in an internally insulated wall structure). The minimum life-cycle cost, USD 2123.06, is for Type-25 (recycled glass fiber as insulation material and pumice stone as building material, in a sandwich wall structure).

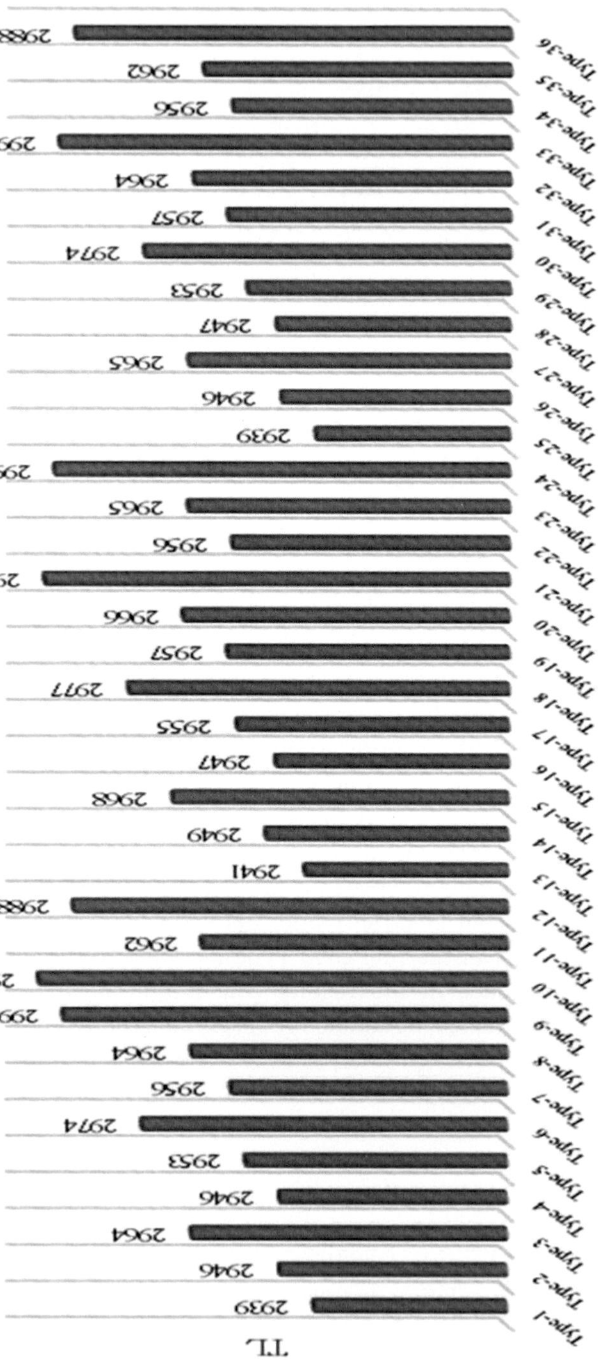

Figure 3.7 Annual energy cost values of harbor-house design scenarios.

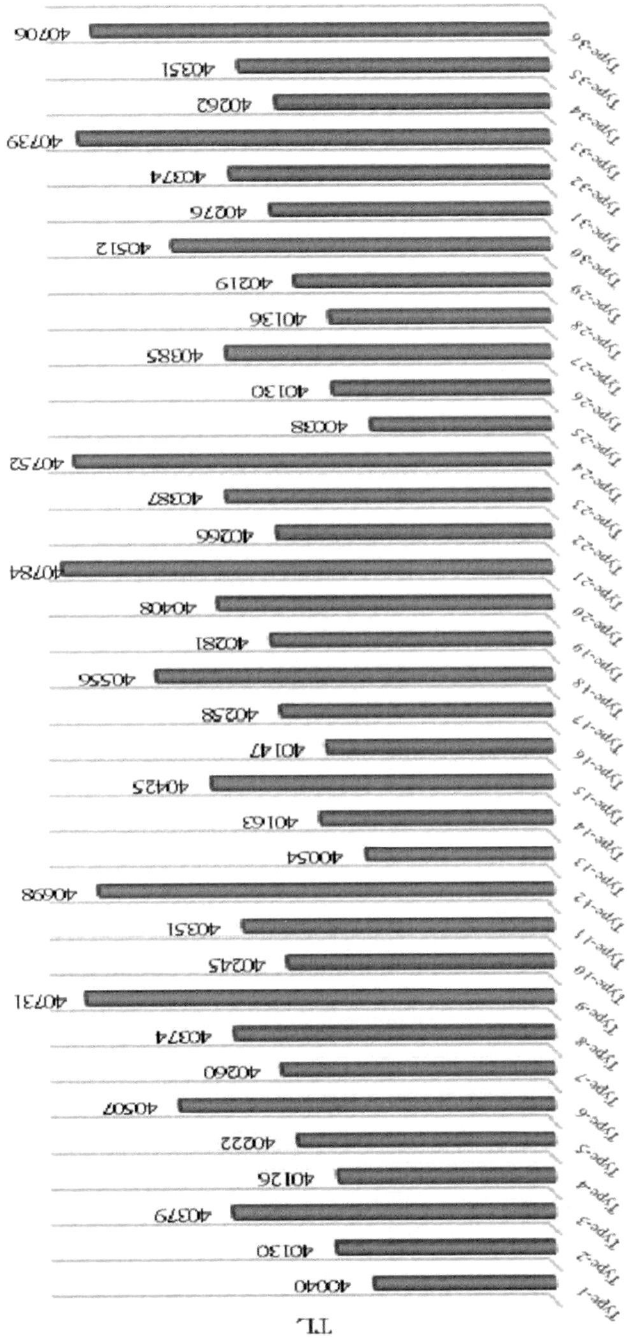

Figure 3.8 Life-cycle costs of designed harbor house.

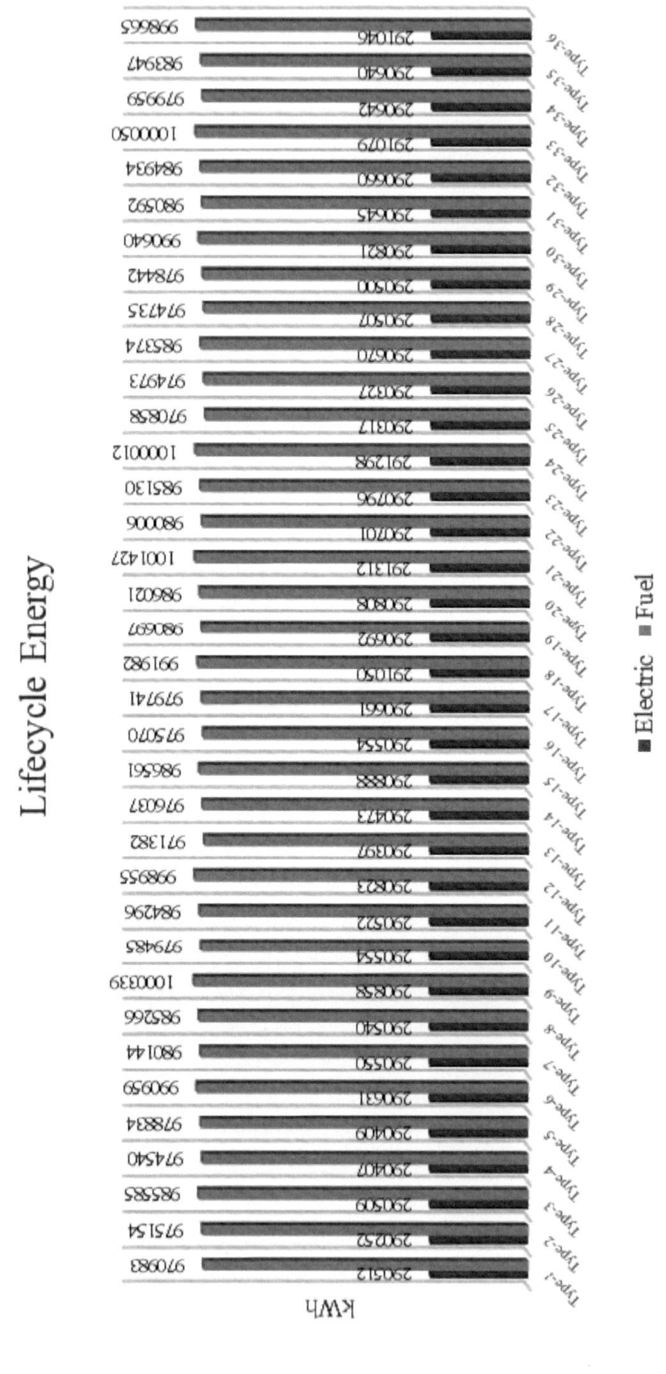

Figure 3.9 Life-cycle electric and fuel values of the harbor-house design scenarios.

Life-cycle electric and fuel values of the harbor-house design scenarios are displayed in Fig. 3.9. The highest life-cycle electric and fuel energy costs are for Type-21 (recycled textile as insulation material and brick as building material, in an internally insulated wall structure), at USD 15447.17 and USD 53101.87, respectively.

The lowest life-cycle electric and fuel energy costs are for Type-2 (with externally insulated wall, using recycled glass fiber as insulation material and Bims block as building material) and Type-25 (with sandwich wall, using recycled glass fiber as insulation material and pumice stone as building material), at USD 15390.96 and USD 51480.92, respectively.

3.4 Conclusions

In this study, a harbor house was designed in Çanakkale, Turkey, with the help of BIM-supported Green Building Studio Connect. The building was visualized with the help of the Lumion program. Within the scope of this study, alternative building and recycled insulation materials used in exterior walls are evaluated in terms of energy consumption by using Green Building Studio simulation. The most energy-efficient planning of the building has been researched by changing the materials that make up the building envelope while simulating the climatic conditions for Çanakkale, Turkey. As a result of the study, it was found that the most efficient wall type is a sandwich wall structure. In the application of 6 cm recycled insulation material on all walls, the most energy-efficient insulation material is recycled glass fiber. The highest-performing building material is pumice stone.

The efficiency of BIM was demonstrated by the energy performance study to convert a green building model of a typical harbor-house model. Through the use of BIM-based Autodesk Revit and Green Building Studio programs, the building envelope components with minimum energy consumption values for the designed building can be determined and the building can be transformed into a green building. The primary purpose of this study is to use recycled insulation materials, that is, healthy and environmentally friendly insulation materials recommended for green buildings in construction. The second purpose is to determine the insulation material that minimizes the total energy consumption of the building among different recycled insulation materials. It is also to determine the most energy-efficient building combination with common building materials and green roofs.

By applying adjustments as and when necessary, improvements to the building model can increase energy efficiency and reduce energy usage. In this study, the contribution of recycled insulation materials to the energy performance of the building was investigated when used to provide thermal insulation on the building walls. In addition to the location of the buildings, the roof and building materials used in the building envelope, the contribution of thermal insulation materials, which is one of the most effective components in thermal insulation, is also critical. Today, there are great efforts to recycle waste materials in many sectors and reuse them in different sectors. In the case of using insulation materials made of recycled materials as thermal insulation material in the building envelope, no simulation studies have been found in the literature on the effects on the energy performance of the building. This study, which was carried out to determine the most energy-performing recycled insulation material with certain roof and building materials in an exemplary climate region, aims to draw the attention of manufacturers to the possibility of an alternative to fossil-based insulation materials if the technical properties of waste-recycled insulation material are further improved. In addition, this study aims to draw the attention of researchers to the usability of different waste materials by showing the effect of recycled insulation materials on energy performance with simulation support.

References

1. Balo, F. (2011). Energy and economic analyses of insulated exterior walls for four different cities in Turkey, *Energy Education Science and Technology Part A: Energy Science and Research*, 26 (2), 175–188.
2. Horta R.M.C. (2012). Sustainable construction of nearly zero-energy buildings, *Faculdade de Ciências e Tecnologia*. https://run.unl.pt/handle/10362/9143, (2012).

3. Cardoso, L.P.S. (2013). Reabilitação sustentável de edifícios Anteriores a 1951 (Sustainable Refurbishment of Buildings Built before 1951), Universidade do Minho, 2013. http://repositorium.sdum.uminho.pt/handle/1822/30877.
4. Kibert, C.J. (1994). Establishing principles and a model for sustainable construction, Sustainable Construction: Proceedings of the First International Conference of CIBTG 16, November 6–9, *Center for Construction and Environment*, M.E. Rinker Sr. School of Building Construction, College of Architecure, University of Florida, Tampa, Florida, 1994 9780964388611, 888.
5. Mateus, R. and Bragança, L. (2011). Sustainability assessment and rating of buildings: developing the methodology SBToolPT–H, *Build. Environ.*, 46, 1962–1971.
6. Haapio, A. and Viitaniemi, P. (2008). A critical review of building environmental assessment tools, *Environ. Impact Assess. Rev.*, 28, 469–482.
7. Bryde, D., Broquetas, M., and Volm, J.M. (2013). The project benefits of building information modeling (BIM), *Int. J. Proj. Manag.*, 31, 971–980.
8. Succar, B. (2009). Building information modeling framework: a research and delivery foundation for industry stakeholders, *Autom. Constr.*, 18, 357–375.
9. Eastman, C.M. (2011). *BIM Handbook: A Guide to Building Information Modeling for Owners, Managers, Designers, Engineers and Contractors*, John Wiley & Sons, ISBN: 9780470541371.
10. Balo, F. and Ulutaş, A. (2023). Energy-performance evaluation with Revit analysis of mathematical-model-based optimal insulation thickness, *Buildings*, 13 (2), 408.
11. Wilkinson, P. (2012). *BIM: An Overview*. https://pt.slideshare.net/EEPaul/cimcig-bimconference, (2012).
12. Azhar, S. (2011). Building information modeling (BIM): trends, benefits, risks, and challenges for the AEC industry, *Leadersh. Manag. Eng.*, 11, 241–252.
13. Azhar, S. and Brown, J. (2009). BIM for sustainability analyses, *Int. J. Constr. Educ. Res.*, 5, 276–292.
14. Ku, K. and Taiebat, M. (2011). BIM experiences and expectations: the Constructors' perspective, *Int. J. Constr. Educ. Res.*, 7, 175–197.
15. Eadie, R., Browne, M., Odeyinka, H., McKeown, C., and McNiff, S. (2013). BIM implementation throughout the UK construction project lifecycle: an analysis, *Autom. Constr.*, 36, 145–151.

16. Solla, M., Ismail, L.H., and Yunus, R. (2016). Investigation on the potential of integrating BIM into green building assessment tools, *ARPN Journal of Engineering and Applied Sciences*, 11(4), 2412–2418.
17. Wong, J.K.W. and Zhou, J. (2015). Enhancing environmental sustainability over building life cycles through green BIM: A review, *Automation in Construction*, 57, 156–165. https://doi.org/10.1016/j.autcon.2015.06.003.
18. Edwards, R.E., Lou, E., Bataw, A., Kamaruzzaman, S.N., and Johnson, C. (2019). Sustainability-led design: Feasibility of incorporating whole-life cycle energy assessment into BIM for refurbishment projects, *Journal of Building Engineering*, 24, 100697. https://doi.org/10.1016/j.jobe.2019.01.027.
19. Lee, S., Tae, S., Roh, S., and Kim, T. (2015). Green template for life cycle assessment of buildings based on building information modeling: focus on embodied environmental impact, *Sustainability*, 7, 16498–16512.
20. Wong, J.K.W. and Zhou, J. (2015). Enhancing environmental sustainability over building life cycles through green bim: a review, *Autom. Constr.*, 57, 156–165.
21. Abanda, F. and Byers, L. (2016). An investigation of the impact of building orientation on energy consumption in a domestic building using emerging bim (building information modeling), *Energy*, 97, 517–527.
22. Peng, C. (2016). Calculation of a building's life cycle carbon emissions based on ecotect and building information modeling, *J. Clean. Prod.*, 112, 453–465.
23. Lu, Y., Le, V.H., and Song, X. (2017). Beyond boundaries: a global use of life cycle inventories for construction materials, *J. Clean. Prod.*, 156, 876–887.
24. Ghaffarian H. A., Zhang, T., Nwadigo, O., GhaffarianHoseini, A., Naismith, N., Tookey, J., and Raahemifar, K. (2017). Application of ND BIM integrated knowledge-based building management system (BIM-IKBMS) for inspecting post-construction energy efficiency, *Renew. Sustain. Energy Rev.*, 72, 935–949.
25. Najjar, M., Figueiredo, K., Palumbo, M., and Haddad, A. (2017). Integration of BIM and LCA: evaluating the environmental impacts of building materials at an early stage of designing a typical office building, *J. Build. Eng.*, 14, 115–126.

26. Xie, J. (2017). Calculation of carbon emissions based on BIM green buildings, *Agro Food Ind. Hi-Tech*, 28, 2526–2529.

27. Marzouk, M., Abdelkader, E., and Al-Gahtani, K. (2017). Building information modeling-based model for calculating direct and indirect emissions in construction projects, *J. Clean. Prod.*, 152, 351–363.

28. Gan, V., Deng, M., Tse, K., Chan, C., Lo, I., and Cheng, J. (2018). Holistic BIM framework for sustainable low carbon design of high-rise buildings, *J. Clean. Prod.*, 195, 1091–1104.

29. Yang, X., Hu, M., Wu, J., and Zhao, B. (2018). Building-information-modeling enabled life cycle assessment, a case study on carbon footprint accounting for a residential building in China, *J. Clean. Prod.*, 183, 729–743.

30. Petrova, E., Pauwels, P., Svidt, K., and Jensen, R. (2019). Towards data-driven sustainable design: decision support based on knowledge discovery in disparate building data, *Archit. Eng. Des. Manag.*, 15, 334–356.

31. Muller, M., Esmanioto, F., Huber, N., Loures, E., and Canciglieri, O. A. (2019). Systematic literature review of interoperability in the green building information modeling lifecycle, *J. Clean. Prod.*, 223, 397–412.

32. Lu, K., Jiang, X., Tam, V., Li, M., Wang, H., Xia, B., and Chen, Q. (2019). Development of a carbon emissions analysis framework using building information modeling and life cycle assessment for the construction of hospital projects, *Sustainability*, 11, 6274.

33. Palumbo, E., Soust-Verdaguer, B., Llatas, C., and Traverso, M. (2020). How to obtain accurate environmental impacts at early design stages in BIM when using environmental product declaration, A Method to Support Decision Making, *Sustainability*, 12, 6927.

34. Wang, N., Satola, D., Wiberg, A., Liu, C., and Gustaysen, A. (2020). Reduction strategies for greenhouse gas emissions from high-speed railway station buildings in a cold climate zone of China, *Sustainability*, 12, 1704.

35. Pucko, Z., Maucec, D., and Suman, N. (2020). Energy and cost analysis of building envelope components using BIM: a systematic approach, *Energies*, 13, 2643.

36. Montiel-Santiago, F., Hermoso-Orzaez, M., and Terrados-Cepeda, J. (2020). Sustainability and energy efficiency: BIM 6D, Study of the

BIM Methodology Applied to Hospital Buildings, Value of Interior Lighting and Daylight in Energy Simulation, *Sustainability*, 12, 5731.

37. Wen, Q., Li, Z., Peng, Y., and Guo, B. (2020). Assessing the effectiveness of building information modeling in developing green buildings from a lifecycle perspective, *Sustainability*, 12, 9988.
38. Kaewunruen, S., Peng, S., and Phil-Ebosie, O. (2020). Digital twin aided sustainability and vulnerability audit for subway stations, *Sustainability*, 12, 7873.
39. Dalla Mora, T., Bolzonello, E., Cavalliere, C., and Peron, F. (2020). Key parameters featuring BIM-LCA integration in buildings: a practical review of the current trends, *Sustainability*, 12, 7182.
40. Fokaides, P.A., Apanaviciene, R., Cerneckiene, J., Jurelionis, A., Klumbyte, E., Kriauciunaite-Neklejonoviene, V., Pupeikis, D., Rekus, D., Sadauskiene, J., and Seduikyte, L. (2020). Research challenges and advancements in the field of sustainable energy technologies in the built environment, *Sustainability*, 12, 8417.
41. Carvalho, J.P., Alecrim, I., Braganca, L., and Mateus, R. (2020). Integrating BIM-based LCA and building sustainability assessment, *Sustainability*, 12, 7468.
42. Shukra, Z.A. and Zhou, Y. (2020). Holistic green BIM: a scientometrics and mixed review, *Eng. Constr. Archit. Manag.*, 28, 2273–2299.
43. Figueiredo, K., Pierott, R., Hammad, A., and Haddad, A. (2021). Sustainable material choice for construction projects: a life cycle sustainability assessment framework based on BIM and Fuzzy-AHP, *Build. Environ.*, 196, 107805.
44. Li, X.-J., Lai, J., Ma, C., and Wang, C. (2021). Using BIM to research carbon footprint during the materialization phase of prefabricated concrete buildings: a China study, *J. Clean. Prod.*, 279, 123454.
45. Kurian, R., Kulkarni, K.S., Ramani, P.V., Meena, C.S., Kumar, A., and Cozzolino, R. (2021). Estimation of carbon footprint of residential building in warm humid climate of India through BIM, *Energies*, 14, 4237.
46. Marzouk, M., Ayman, R., Alwan, Z., and Elshaboury, N. (2022). Green building system integration into project delivery utilising BIM, *Environ. Dev. Sustain.*, 24, 6467–6480.
47. Gardezi, S., Shafiq, N., and Khan, M. (2021). Relational pre-impact assessment of conventional housing features and carbon footprint for achieving sustainable built environment, *Environ. Dev. Sustain.*, 24, 8441–8463.

48. Thormark. C. (2006). The effect of material choice on the total energy need and recycling potential of a building, *Building and Environment*, 41, 10191026.

49. Mutani, G., Pastorelli M., and de Bosio, F. (2015). A model for the evaluation of thermal and electric energy consumptions in residential buildings, *4th International Conference on Renewable Energy Research and Applications*, Palenno, Italy, 1399–1404, 22–25 Nov 2015.

50. Matuampunwong, D., Chirarattananon, S., and Chaiwiwatworakul, P. (2012). Energy saving potential from daylighting through external multiple-slat shaded window in the tropics, *International Journal of Renewable Energy Research*, 2 (3), 376–383.

51. Ministry of New and Renewable Energy, India (2012). Thermal performance of buildings. Available at http://mnre.gov.in/solar-energy/ch4.pdf.

52. Imran M. and Mughal, H. U. (2013). Energy analysis of roof integrated solar collector for domestic heating & cooling under local conditions of Pakistan, *International Journal of Renewable Energy Research*, 3 (1), 14–19.

53. Kürekçi, A. Bardakçı, A.T., Çubuk, H., and Emanet, Ö. (2012). Türkiyénin Tüm İlleri İçin Optimum Yalıtım Kalınlığının Belirlenmesi, Tesisat Mühendisliği–No. 131 – Sep./Oct. 2012, 5–21.

Chapter 4

Recent Progress in Photovoltaic-Thermoelectric Hybrid Power Systems

Xi Wang, Paul Henshaw, and David S.-K. Ting

Turbulence and Energy Laboratory, University of Windsor, Ontario, Canada

wang1st@uwindsor.ca

Photovoltaic-thermoelectric (PV-TE) is a novel hybrid electricity generation system that has the potential to significantly further renewable energy. Reported improvements in solar efficiency of PV-TE systems over PV systems range from 1% to 30% depending on whether the study was experimental or theoretical, and the assumptions used. A PV-TE system's performance is sensitive to the working environments, such as temperature and light. Besides, the geometric structure plays an important role in thermoelectric conversion. Therefore, it is important to improve the hybrid scheme for a PV-TE system based on these characteristics. This chapter introduces some improved PV-TE hybrid schemes from the three aspects: optical design, heat transfer, and geometric structure. Most optical design improvements focus on splitting the incoming solar spectrum, directing the PV-active wavelengths to the PV cells and the long-wave energy to the TEG. Other schemes

Reclaiming Eden: Responsible Living, Engineering, and Architectures
Edited by David S.-K. Ting and Jacqueline A. Stagner

ISBN 978-981-5129-31-1 (Hardcover), 978-1-003-49612-0 (eBook)
www.jennystanford.com

extract heat from the PV cells and direct it to a bank of TEGs, smaller in area than the PV modules. The TEG efficiency itself can also be improved by optimizing the geometry of the N and P semi-conductors. Moreover, except for the performance in energy conversion, cost-effectiveness is a significant index to evaluate the prospect of a renewable energy solution. Thus, the economic analysis for a PV-TE system is reviewed in this paper. Overall, the main purpose is to serve as a useful reference for future studies of PV-TE hybrid systems.

Nomenclature

A_{cell}	PV cell area
a	Diode's ideality factor
B	Future values of the benefit
C	Future values of the cost
G_t	Solar radiation
I_{Ph}	Photocurrent
I_0	Reverse saturation current of the diode
I_{TEG}	Current produced by a TEG
j	Time period
K	Boltzmann constant
$\bar{K}$	Net thermal conductance
MCHP	Micro-channel heat pipe
N_{PV}	Number of PV panels
NPV	Net present value
P_{hybrid}	Total power produced by A PV-TE hybrid system
PV	Photovoltaic
PV-TE/T	Photovoltaic-thermoelectric/thermal
PCM	Phase change material
P_{PV}	Output power produced by a PV cell
P_{TEG}	Output power produced by a TEG
$\dot{Q}_F$	Fourier heat transfer rate

$\dot{Q}_{input}$	Input heat transfer rate
$\dot{Q}_J$	Joule heat transfer rate
$\dot{Q}_P$	Peltier heat transfer rate
r	Interest rate
R_i	Internal resistance
R_L	Load resistance
S	Seebeck coefficient
TE	Thermoelectric
TEG	Thermoelectric generator
T_c	Temperature of the cold side
T_h	Temperature of the hot side
V_T	Temperature voltage
V_{TEG}	Voltage produced by a TEG
η_{hybrid}	Overall efficiency of a PV-TE hybrid system
η_{PV}	Efficiency of a PV cell
η_{TEG}	Efficiency of a TEG

4.1 Introduction

Greenhouse gas emissions are considered as the main reason for global warming, thereby having become the focus of the public's concerns. Many countries, such as European countries and Japan, have made commitments to achieve carbon neutrality in 2050 [1]. However, the global energy demand has risen due to the economic recovery and energy market conditions during the post-COVID-19 pandemic period, causing many traditional power plants to be recommissioned. In 2021, the global CO_2 produced from industrial activities and combustion processes rebounded to the highest-ever annual level. According to the International Energy Agency (IEA), the world must make sure that this global emission rebound was a one-off, in order to keep on track for carbon neutrality in 2050 [2]. At the current stage, it is necessary to intensify the investment in renewable energy technology, making it possible to replace the carbon-intensive economy further.

As one of the important renewable energy technologies, solar power has been developing at an astounding pace in recent years. In 2021, there was a record increase in clean electricity generation to 522 TWh, which is 7% of the total. Of this, the electrical power generated by solar photovoltaic (PV) takes up almost one-third of the generation growth [3].

The astounding development in solar power technology has a positive effect on decarbonization. Chen et al. [4] reported a decrease of 0.53 kg in CO_2 emission for every kWh of electrical power generated by a solar PV system. It is also estimated that CO_2 emissions can be reduced by 6.5% to 18.8% in the USA when 10% of electricity is produced by PV systems [5]. Furthermore, by 2030, PV technology may cause a decrease in CO_2 emissions by about 69–100 million tons [6]. The "carbon footprint" is used to evaluate the greenhouse gas emissions due to human activities. The carbon footprint of a PV system ranges from 14 g to 73 g CO_2-eq/kWh, which is far less than that of fossil-fuel-generated electricity, for which the value ranges between 607.6 and 975.3 CO_2-eq/kWh [6]. Therefore, PV technology can play an important role in alleviating global warming.

Although PV technology is a proven green energy, further advancement can be realized from the temperature angle. It is known that PV cells cannot convert radiant energy from the whole solar spectrum (ranging from 280 nm to 4000 nm) to electricity [7]. Those absorbed but unused photons tend to increase the temperature of the PV panel, causing its energy conversion efficiency to drop. Typically, a temperature increase of 1 °C may cause an efficiency decrease of 0.08–0.45% [8–10]. Although an efficient cooling system is one way of dealing with this challenge, it is more attractive to develop novel schemes to recover the wasted heat generated from a PV system. Among many schemes, the thermoelectric generator (TEG) is considered as one of the ideal thermal recovery technologies for a PV system. The TEG is a semiconductor-based energy converter driven by temperature difference [11]. The TEG is equipped with a compact structure and does not produce any extra emissions during the process of thermal recovery [11]. In this way, an increasing number of researchers are paying more attention to the PV-TE hybrid system, recently.

In a traditional PV-TE hybrid system, the wasted heat produced by a PV cell can be transferred via its back to the hot side of a TEG module, making it possible to form a temperature difference across the TEG's two sides. The PV cell in a hybrid system can produce electrical power utilizing the photovoltaic effect. Meanwhile, the temperature difference can drive a TEG module to produce power [12]. In a simplistic case, it could be assumed that the change in heat loss from the PV, as a result of adding TEG on the back, is negligible [9]. In actual fact, adding TEGs may increase the temperature of the PV panel, which has a negative impact on PV work [13, 14]. Pang et al. [14] utilized an experiment to verify that the temperature in a PV-TE hybrid system is 1.8 °C higher than the PV-only with a heat sink as an add-on under a natural convection condition. However, many studies have indicated that the performance of a PV-TE hybrid system is normally better than that of a PV-only system because the extra power generated by a TEG module can compensate for the negative effect of the temperature on a PV panel to some extent [13, 15]. According to the studies by Van Sark [16] in Malaga, Spain, and Utrecht, the Netherlands, it was found that the annual energy yield of a PV roof increases by 14.7–11% through adding TEG modules. Generally, despite that the energy feasibility of a PV-TE system has been demonstrated, many researchers are keen to design or optimize the hybrid scheme further, based on its operating characteristics [17].

Firstly, as for a PV-TE hybrid system, the irradiance has a direct effect on the performance of a PV cell, while indirectly influencing the TEG module. It is worth noting that the spectral response differs for PV materials. In this way, some hybrid systems with a beam splitter were proposed, which can reflect the matched solar spectrums on a specific PV cell, and transmit others to maximize the TEG performance [18, 19]. Secondly, the effects of the working temperature on the PV cell and TEG module in a hybrid system are different. The energy conversion of a PV cell can be decreased by increased temperature [8–10]. However, the thermoelectric conversion is generally increased by the temperature across the device. There is an optimal temperature for a TEG with a specific material [20, 21]. Some researchers tried to improve the cooling system for a PV-TE scheme, making it possible to ensure the PV cell's performance and match an appropriate temperature for the TEG simultaneously. Besides, the geometric structure plays a significant role in a TEG's performance [21–23]. Therefore, it is worthwhile to propose some novel structures, such as the segment or variable cross-sectional designs, for a TEG, which can make full use of the temperature driving force in a hybrid system.

A review of the latest advancements in PV-TE hybrid systems is presented in this chapter. Through a brief introduction to the photovoltaic effect and thermoelectric effect, the basic working principle of a PV-TE system is illustrated. Then, some improved schemes of hybrid systems are reported, which mainly involve three different improvements: optical system, cooling system, and geometric structure. Overall, this chapter provides readers with a useful summary of the design and optimization research of PV-TE hybrid systems.

4.2 Working Principle of a PV Cell

The basic principle of a PV cell is the photoelectric effect, which can convert solar energy to electricity through the transition of electrons in a semiconductor under light conditions. Figure 4.1 shows the schematic diagram of a PV cell unit. A PV cell usually consists of P and N semiconductor layers. In the semiconductor, its valence electrons are easily excited under light conditions [24]. When the energy of photos in the light exceeds that of the band gap, the excited electrons in the semiconductor will transit to the conduction band and become free electrons. Under an internal electric field in the semiconductor layers, these free electrons are available to form current through an external circuit, as shown in Fig. 4.1.

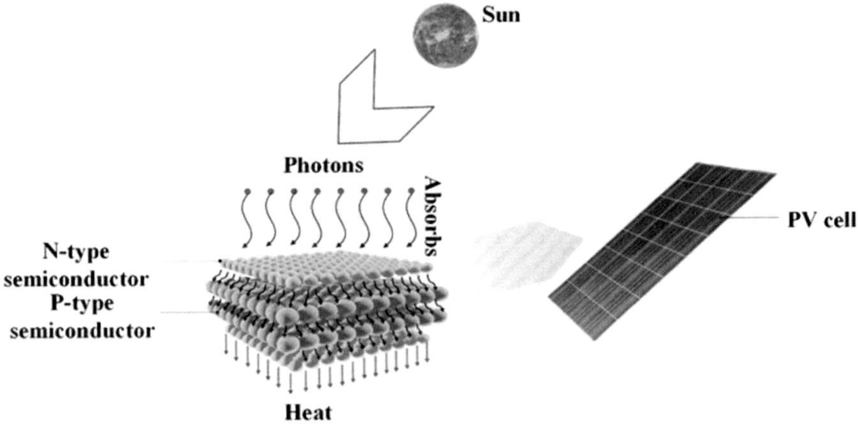

Figure 4.1 The working principle of a PV cell. Recreated based on [7].

The power generated (P_{pv}) by a PV cell is the product between the current (I) through and voltage (V) across the load resistance, and the efficiency (η_{pv}) of a PV cell is a ratio of the output power and incident light power [25],

$$P_{pv} = VI \tag{4.1}$$

$$\eta_{pv} = \frac{P_{pv}}{A_{cell}G_t}. \tag{4.2}$$

In order to evaluate a PV cell based on Eqs. (4.1) and (4.2), it is necessary to establish an accurate output characteristic curve of (I vs V). Normally, a PV cell can be modeled as a series resistance (R_s), a shunt resistance (R_{sh}), and diodes [26, 27]. The R_s is actually the resistance inside each cell, and R_{sh} represents the internal shunt resistance of diodes. Therefore, the common PV cell model is the single diode model (SDM). In the SDM model, there is one diode connecting with the series resistance and a shunt resistor in parallel, as shown in Fig. 4.2. The I–V characteristics in the SDM model can be expressed as,

$$I = I_{ph} - I_0\left[\exp\left(\frac{q(V+IR_s)}{aKT} - 1\right)\right] - \frac{V+IR_s}{R_{sh}}, \tag{4.3}$$

where a is the diode's ideality factor. T is the PV cell temperature. The Boltzmann constant is symbolled by K, and its value is 1.38×10^{-23} J/K [27]. I_{ph} and I_0 represent the photocurrent and reverse saturation current of the diode, respectively.

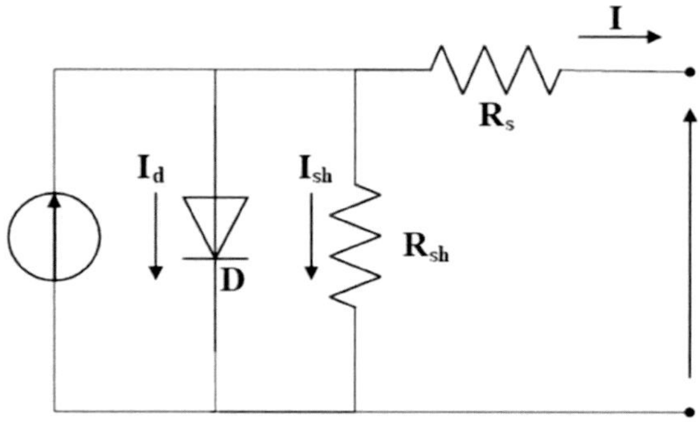

Figure 4.2 Equivalent circuit diagram of the SDM model. Recreated based on [26].

The SDM is widely used to evaluate a PV cell's performance, due to its simple circuit structure and easy hardware implementation [28]. However, it is difficult for this model to describe the output characteristic in the I–V of a PV cell accurately.

4.3 Working Principle of a TEG

In addition to the light, the valence electrons in a semiconductor can be excited by thermal energy to form a current, this is called the Seebeck effect. Specifically, the Seebeck effect describes a phenomenon where the holes in the P-type semiconductor and the electrons in the N-type semiconductor are driven from the hot side to the cold side. This creates an electric potential. According to Fig. 4.3, there are three main components in a TEG module, namely, P/N semiconductors, conductors, and ceramic substrates [29].

The Seebeck effect discloses that the Seebeck voltage (V_{TEG}) of a TEG is directly proportional to the temperature difference between the hot and cold sides, i.e. [29–32],

$$V_{TEG} = S\,(T_h - T_c) \tag{4.4}$$

where the Seebeck coefficient is denoted by S; and T_h and T_c are the temperature at the hot and cold side of a TEG module, respectively.

The TEG module can generate electrical power when connecting a load resistance. Therefore, based on Ohm's law, the current passing through the load resistance (I_{TEG}) is the ratio between the voltage produced by a TEG and the total resistance, as shown in Eq. (4.5). Besides, the output power of a TEG (P_{TEG}) is proportional to the current squared and the load resistance (as shown in Eq. (4.6)) [30, 31],

$$I_{TEG} = \frac{V_{TEG}}{R_{total}} = \frac{S(T_h - T_c)}{(R_i + R_L)} \tag{4.5}$$

$$P_{TEG} = I_{TEG}^2 R_L = \left[\frac{S(T_h - T_c)}{(R_i + R_L)} \right]^2 R_L, \tag{4.6}$$

in which, the total resistance, load resistance, and internal resistance are named R_{total}, R_L, and R_i, respectively.

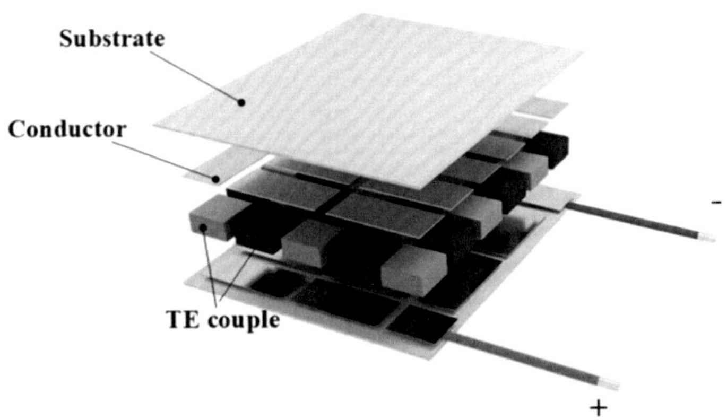

Figure 4.3 The schematic diagram of a TEG module.

Heat transfer is the underlying mechanism for TEG operation. Thus, it is necessary to make clear the thermal process in order to evaluate the efficiency of a TEG module. According to the heat transfer process in a TEG (refer to Fig. 4.4), four kinds of thermal energy (energy input, Joule heat, Fourier heat, and Peltier heat) pass through the hot side substrate and conductor [29]. The energy input and half of Joule heat are absorbed by the substrate while the Fourier heat and Peltier heat are released from the boundary [32].

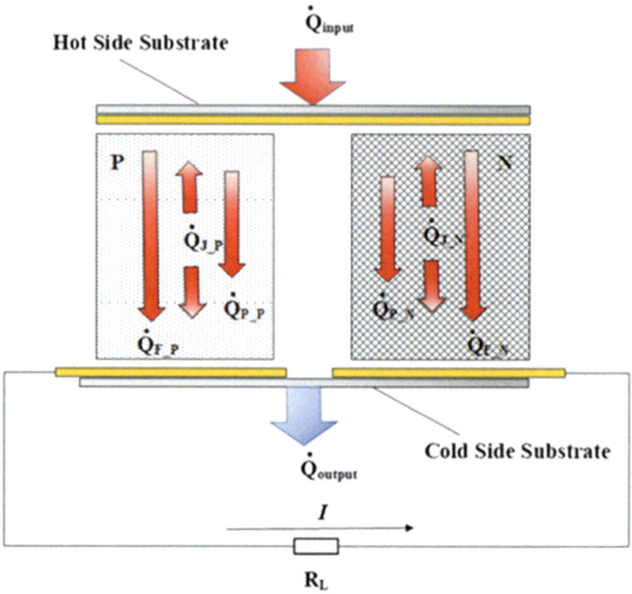

Figure 4.4 The heat transfer process in the hot side thermal boundary of a TEG.

According to energy conservation, the energy input rate ($\dot{Q}_{in}$) is the algebraic sum of Fourier heat rate ($\dot{Q}_F$), Peltier heat rate ($\dot{Q}_P$), and Joule heat ($\dot{Q}_J$), as shown in Eq. (4.7) [32],

$$\dot{Q}_{in} = \dot{Q}_F + \dot{Q}_P - \frac{1}{2}\dot{Q}_J. \tag{4.7}$$

In the energy equation, the Fourier heat transfer rate is decided by the temperature difference and net thermal conductance, i.e. [32],

$$\dot{Q}_F = \bar{K}(T_h - T_c). \tag{4.8}$$

When a current passes through a TEG, a part of electricity will be converted to thermal energy due to its internal resistance, which is the Joule heat. The Joule heat rate is proportional to the current square and the internal resistance [33], as shown in Eq. (4.9).

$$\dot{Q}_J = I_{TEG}^2 R_i = \left[\frac{S(T_h - T_c)}{R_i + R_L}\right]^2 R_i. \tag{4.9}$$

According to the Peltier law, a cooling or heating effect can be produced when a current passes through a P/N junction. In this way, the heat produced by the Peltier effect is related to the current and temperature of the P/N junction [24]. The Peltier heat on the hot side can be calculated by Eq. (4.10) [29],

$$\dot{Q}_P = ST_h I_{TEG} = \frac{S^2 T_h (T_h - T_c)}{R_i + R_L}. \tag{4.10}$$

Overall, the efficiency (η_{TEG}) of a TEG module is a ratio between the output power and input heat rate, as shown in Eq. (4.11),

$$\eta_{TEG} = \frac{\left[\frac{S(T_h - T_c)}{R_i + R_L}\right]^2 R_L}{\bar{K}(T_h - T_c) + \frac{S^2 T_h (T_h - T_c)}{R_i + R_L} + \frac{1}{2}\left[\frac{S(T_h - T_c)}{R_i + R_L}\right]^2 R_i}. \tag{4.11}$$

4.4 Traditional PV-TE Hybrid System

Currently, PV cells fail to convert all photons from the sun to electricity. Normally, the absorbed photons with wavelengths near the infrared region are converted to heat, which has negative impacts on a PV cell's performance. However, the thermal energy produced from a PV cell can be used to drive a TEG module. In this way, the TEG is considered as a feasible way to recover the wasted heat of a PV cell, making it possible to improve the overall performance of the energy system.

In a traditional PV-TE hybrid scheme (which is a simple integrated hybrid scheme), the TEG module is sandwiched between a PV cell and a radiator (Fig. 4.5) [11]. The front of the

PV cell is used to receive solar radiation. Meanwhile, the wasted heat is transferred from the back of the PV cell to the hot side of the TEG module, passes through the module, and is dissipated to the surroundings by the cooling devices at its cold side. In this way, the TEG can be driven by the temperature difference between its two sides to generate electrical power.

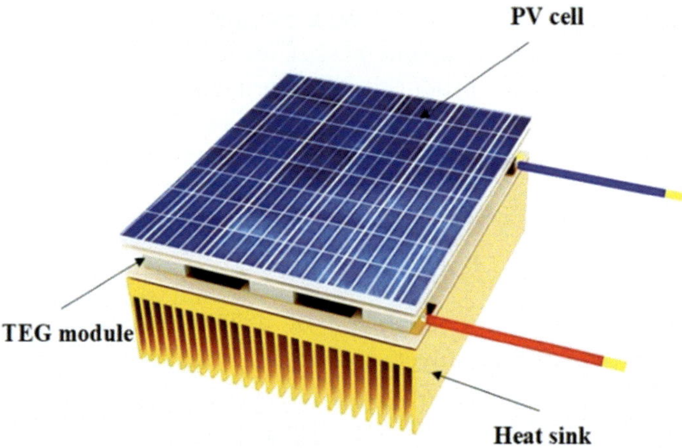

Figure 4.5 The schematic diagram of a simple PV-TE hybrid system.

The total power (P_{hybrid}) produced by the hybrid system can be calculated by Eq. (4.12) [34], and the overall efficiency (η_{hybrid}) of the hybrid system is the ratio between the total power and incident light power, as shown in Eq. (4.13) [35],

$$P_{hybrid} = P_{pv} + P_{TEG} \tag{4.12}$$

$$\eta_{hybrid} = \frac{P_{hybrid}}{A_{cell} G_t} = \frac{P_{pv} + P_{TEG}}{A_{cell} G_t}. \tag{4.13}$$

4.5 Improved Hybrid Schemes of the PV-TE System

The performance of some representative PV-TE hybrid systems has been discussed in previous sections. Liwanag et al. [36]

utilized an experiment to verify that there is a 1–1.9% increase in the average power output generated by a PV-TE hybrid system compared to the PV-only system. Khan et al. [37] experimentally tested a polycrystalline silicon-based PV-TE hybrid system at Lahore (31.5204° N, 74.3587° E), Pakistan. They collected averaged data of the hybrid PV-TE and PV-only systems for 10 consecutive days in winter. The results (refer to Fig. 4.6) indicate that integrating TEGs is an effective measure to enhance the performance of PV systems. The improvement in the output power reaches the maximum between 11 am and 12 noon. During this period, the efficiency of the hybrid PV-TE system is 14%, which is higher than that of the PV-only system (11.6%).

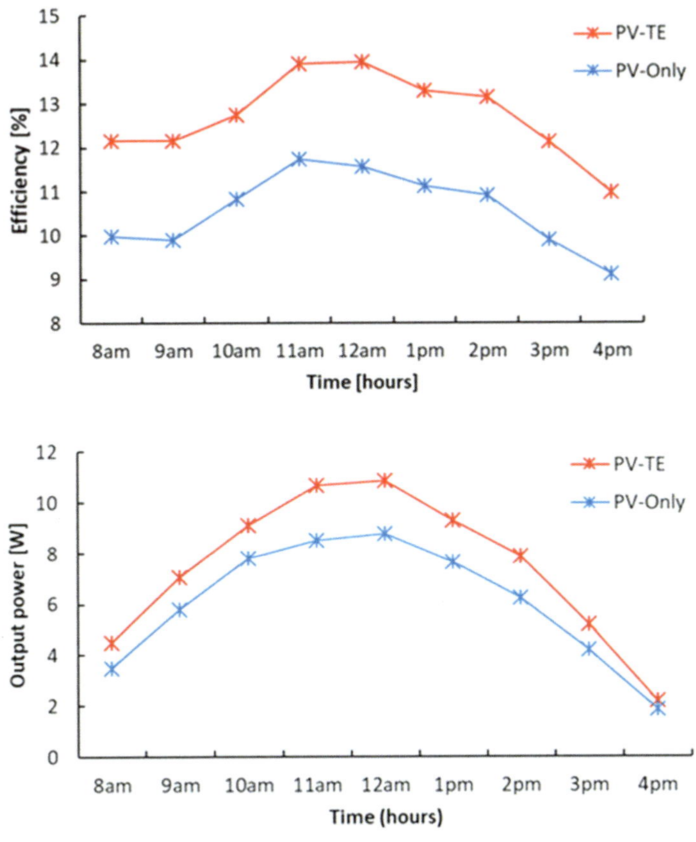

Figure 4.6 Transient performances of the PV-only and hybrid PV-TE systems: (a) output power and (b) efficiency. Recreated based on [37].

Besides, Wehbi et al. [15] reviewed some hybrid PV-TE systems and compared their overall efficiency with stand-alone PV systems. According to Fig. 4.7, the overall efficiency of a hybrid PV-TE system is higher than that of a PV-only system. The enhancement in overall efficiency even can exceed 30%.

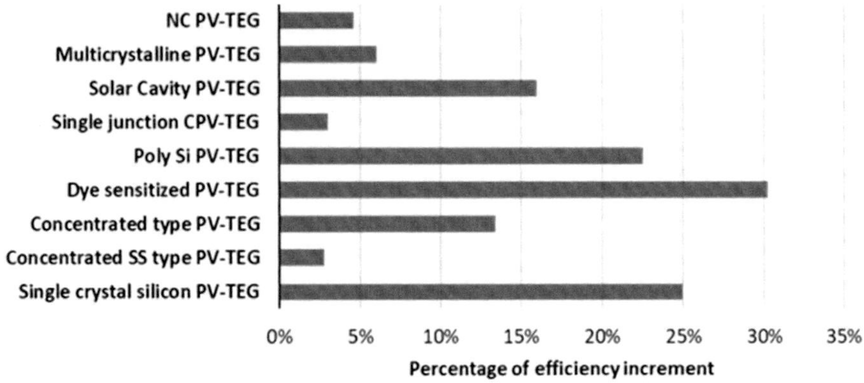

Figure 4.7 Efficiency enhancement of a hybrid PV-TEG system compared to a PV-only system. Recreated based on [15].

However, the performance of a traditional PV-TE hybrid system is far from optimal, which has been disclosed in some studies. According to the study of Liwanag et al. [36], the enhancement of the average power output can reach 27.9% in theory by adding TEG modules to the back of a PV system. Actually, there is a huge gap between experimental and theoretical results. A similar problem can also be found in the research of Van Sark [13]. In order to make a PV cell and TEG work at optimal states, it is necessary to match appropriate working conditions (like light and temperature, etc.) and geometric structures (such as TE couple shapes) for them. There is enormous potential to improve a PV-TE hybrid scheme. Recently, some researchers strived to improve the hybrid scheme of a PV-TE system through three aspects: optical design, heat transfer, and geometric structure. Table 4.1 displays the main studies in PV-TE hybrid system improvement.

Table 4.1 Summary of recent studies in PV-TE system improvement

Author	Technique	Contribution
Shou et al. [38]	Spectrum splitter	Designed an 82-layer optical interference thin-film filter consisting of TiO_2/SiO_2 to increase an average efficiency of 3.24% for a PV-TE system
Mahmoudinezhad et al. [39]	Spectrum splitter	It was verified by an experiment that the splitting PV-TE system can output more power compared to the PV-only system
Alnajideen and Min [40]	Spectrum splitter	Designed four rectangular dichroic mirrors for a PV-TE hybrid system to improve overall efficiency by 6.3%
Zhang et al. [41]	Thermal interface materials	Utilized thermal materials to increase power of 14% and 60% generated by the PV cell and TEG in a hybrid system, respectively
Makki et al. [42]	Heat pipe	Designed a heat pipe to increase the overall performance of a PV-TE system
Shittu et al. [43]	Flat plate heat pipe	Utilized a flat plate heat pipe to improve the overall efficiency of a PV-TE hybrid system by 3.31% under a concentration ratio of 6
Gao et al. [44]	Heat pipe and Fresnel lens	Verified that a bifacial-PV-TE-system (1.82 W) can output more power than that of a tandem-PV-TE system (1.45 W)
Motiei et al. [45] and Luo et al. [46]	PCM	Verified an appropriate PCM may have a positive effect on the overall performance of a PV-TE hybrid system
Zhang and Xuan [47]	Layout optimization	Designed a V-type groove layout of a PV-TE hybrid system to increase the efficiency by 20% compared to the flat type of system
Cui et al. [48]	Segmented TEG	The efficiency of a segmented PV-TE system is better than that of the traditional hybrid scheme.
Maduabuchi [49]	Variable cross-sectional TEG	Compared to the traditional hybrid system, the variable cross-sectional TEG is preferable for a PV-TE system under insufficient solar irradiance

4.5.1 PV-TE System with a Splitter

There is a wide spectrum of wavelengths in solar radiation, ranging from 280 nm to 4000 nm [7]. The PV cell cannot convert the entire range of wavelengths into electricity. The unconverted photos will cause an increase in the temperature at the panel, which has a negative impact on the performance of the PV cell. The usable range of the spectrum may vary across PV materials. Figure 4.8 shows the spectral response of some PV materials. It is worthwhile to separate the specific spectrum ranges from the incoming radiation, making the PV cell receive only the photos with suitable wavelengths. Meanwhile, other photos are transmitted to the TEG as a heat source. Based on this idea, some researchers proposed a beam splitter for a PV-TE hybrid system to maximize its performance. Differing from the traditional hybrid system, the PV cell and TEG module are placed separately in a spectrum-splitting PV-TE system.

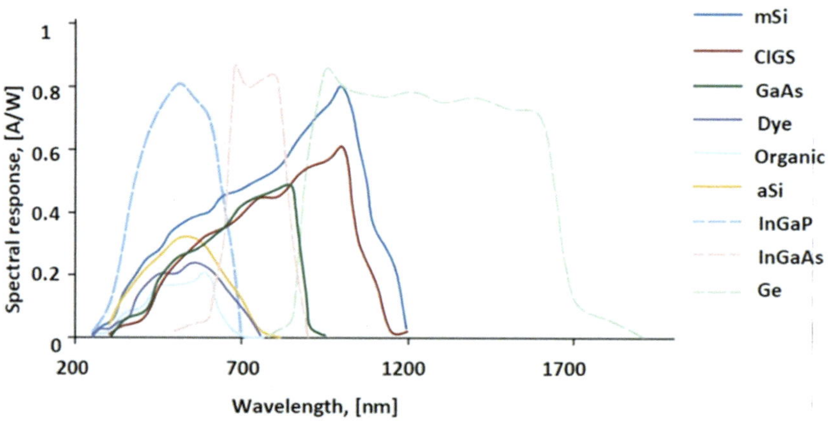

Figure 4.8 Spectral response for several photovoltaic cells. Recreated based on [50].

Shou et al. [38] designed an 82-layer optical interference thin-film filter consisting of TiO_2/SiO_2 for a PV-TE hybrid system, as shown in Fig. 4.9. In their hybrid scheme, the optical splitter can reflect from 400 nm to 1100 nm spectrum for the PV cell. Meanwhile, the solar spectrum from 1100 nm to 2500 nm is transmitted to the TEG module by this filter. The tests under four incidence angles verify that the conversion efficiency of the hybrid system can be improved by the thin-film filter on a splitter.

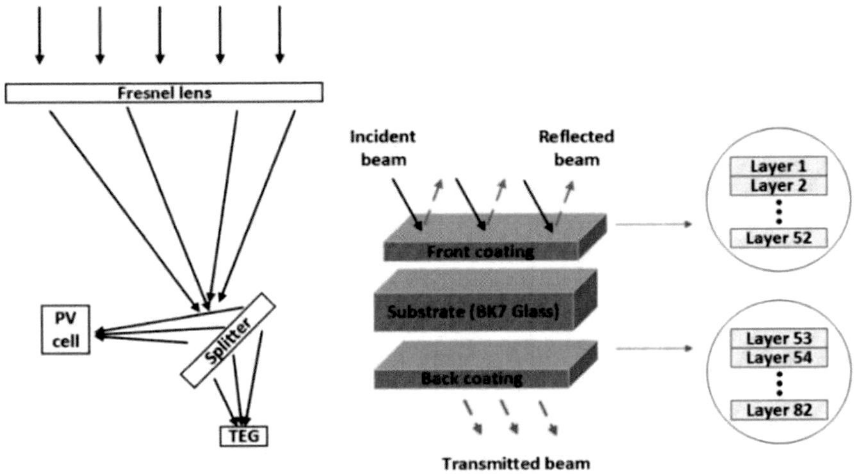

Figure 4.9 Structural diagrams of the PV-TE system and its beam splitter. Recreated based on the work of Shou et al. [38].

An experiment was conducted by Mahmoudinezhad et al. [39] to verify further that the optical splitter is an effective measure to increase the performance of a PV-TE system. In the experiment (as shown in Fig. 4.10), they compared the output performance of three cases: TEG-only system, PV-only system, and PV-TEG-BS hybrid system. The results indicate that compared to the TEG-only system, the splitter may decrease the power generation of the TEG module. However, the splitting technology can improve the output power dramatically for the PV cell in the hybrid system. Besides, the output power of the PV cell in the hybrid system is higher than that of the PV-only system. Therefore, the splitting technology has a positive effect on the overall performance of the PV-TE hybrid system.

Additionally, Alnajideen and Min [40] set four rectangular dichroic mirrors in a PV-TE system to reflect and concentrate the visible spectrum to the PV cell; meanwhile, the infrared spectrum was transmitted to a heat absorber, which is used to drive the TEG module. The schematic diagram of the novel hybrid PV-TE system is displayed in Fig. 4.11. The tested results verify that there is a 6.3% improvement in the efficiency of the novel hybrid system.

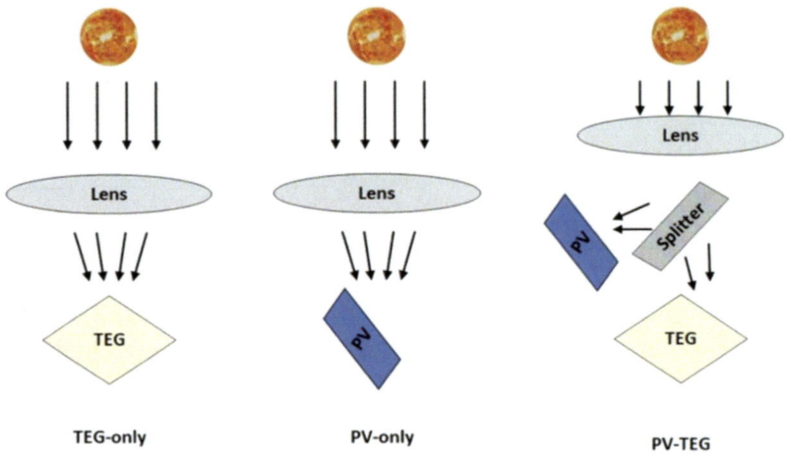

Figure 4.10 Three experimental cases in the study of Mahmoudinezhad et al. [39]

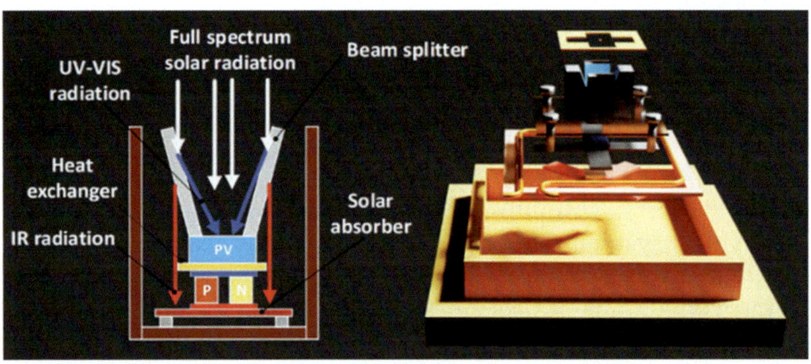

Figure 4.11 Structural diagrams of the PV-TE system and its beam splitter. Recreated based on [40].

4.5.2 PV-TE System with an Advanced Heat Transfer Technology

The heat transfer process plays an important role in a PV-TE hybrid system. Theoretically, it is desirable to decrease the temperature of the PV panel and enlarge the temperature difference for the TEG as far as possible. Some advanced heat transfer technologies have been applied in a PV-TE hybrid system. Firstly, in a PV-TE hybrid system, there are considerable thermal resistances at the junctions of different components, which has a negative impact on the performance of the system to some extent. Zhang et al. [41] introduced a thermal interface material to fill the gap between the PV cell and TEG, as shown in Fig. 4.12. They verified that by using the interface material to decrease the contact thermal resistance, the power generation of the PV cell and TEG in the hybrid system can be increased by 14% and 60%, respectively.

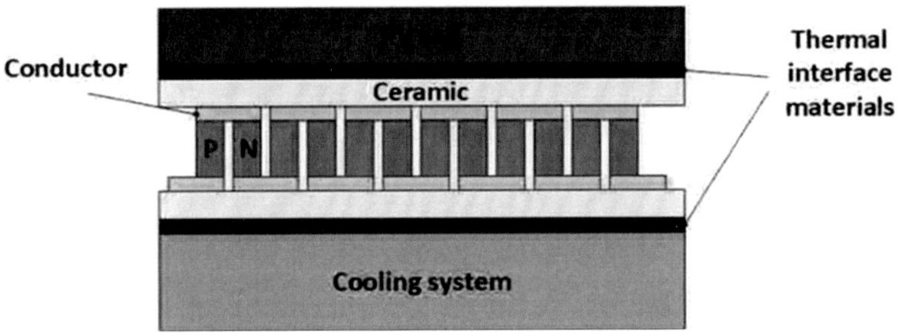

Figure 4.12 Layout of the thermal interface materials in a PV-TE system. Recreated based on [41].

Heat pipes use phase-changing to efficiently convey heat from hot areas to cold areas. Based on this principle, many heat pipes were designed and applied to improve the heat transfer process for a PV-TE hybrid system. Makki et al. [42] designed a PV-TE hybrid scheme in which a heat pipe carried heat between the PV cells and TEG modules (as shown in Fig. 4.13). They found the heat pipe can transfer the wasted heat from the PV cells to the TEG modules effectively, making it possible to improve the overall performance of the hybrid system.

Figure 4.13 Structure diagram of the PV-TE heat pipe system. Recreated based on [42].

Utilizing a numerical method, Shittu et al. [43] also verified that a flat plate heat pipe (as shown in Fig. 4.14) is an effective way to improve the performance of a PV-TE hybrid system. They established three numerical models: PV-only, PV-TE, and PV-TE-HP. Through the comparison of the performance of the three power systems, it was found that the efficiency of the PV-TE-HP system tops the group. Its efficiency is 58.01% and 3.31% higher than that of the PV-only and PV-TE systems.

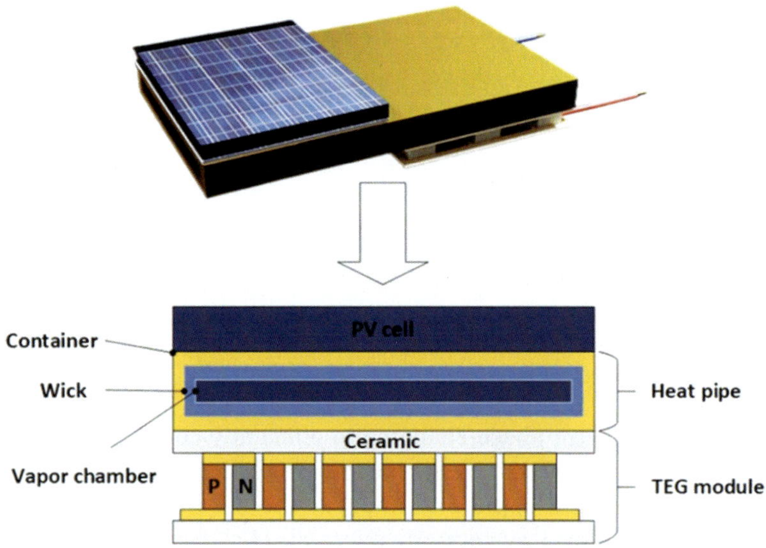

Figure 4.14 A PV-TE system with a flat plate heat pipe. Recreated based on [43].

Gao et al. [44] proposed two kinds of hybrid systems with a heat pipe, namely, the bifacial-PV-TE-system and the tandem-PV-TE system, as shown in Fig. 4.15. They found the power generated by the bifacial-PV-TE-system (1.82 W) is about 25.1% higher than that of the tandem-PV-TE system (1.45 W) under the solar irradiance with 10000 W/m^2 and ambient temperature with 298.15 K.

In addition, a phase change material (PCM) is considered as an excellent energy storage medium because it can absorb or emit a large amount of latent heat during a phase-changing process.

An appropriate PCM is also a useful way to reduce the temperature at the PV panel and enlarge the mean temperature difference of the TEG in a transient application scenario. Based on the layout of the PCM in a hybrid system, there are two kinds of schemes: PV-TEG-PCM and PV-PCM-TEG, as shown in Fig. 4.16. Motiei et al. [45] and Luo et al. [46] utilized numerical methods to verify that both these two kinds of improved schemes resulted in better performance than that of the traditional hybrid system. Luo et al. [46], also made simulations to compare the performance of PV-PCM-TE, PV-TE, and PV-only systems year-round. It is verified that the PCM can be used to enhance the hybrid PV-TE system in the long term, due to its storage and release of heat. The total electrical work and average efficiency of the PV-PCM-TE system are higher than the values for the PV-only and PV-TE systems by 10.15% and 2.37%, respectively.

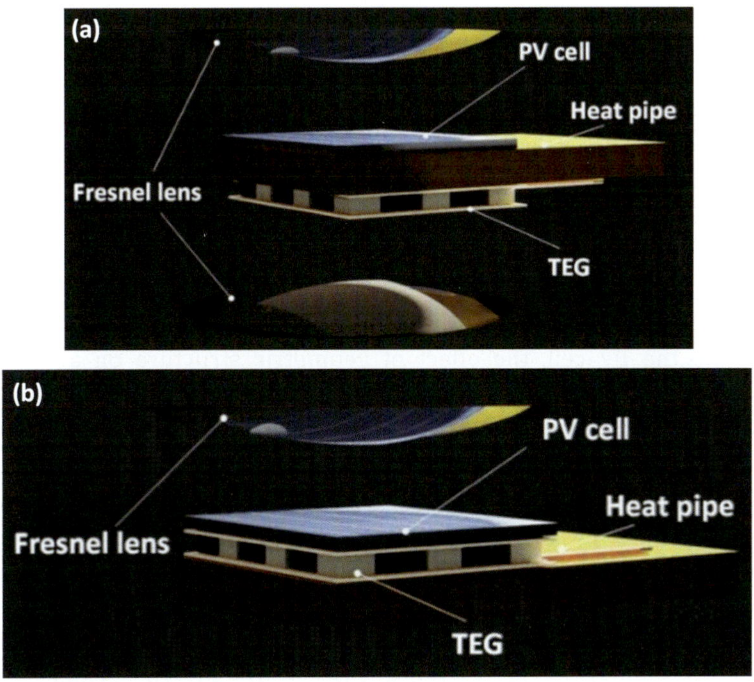

Figure 4.15 Two kinds of layouts of the PV-TE heat pipe system: (a) bifacial-PV-TE-system and (b) tandem-PV-TE system. Recreated based on [44].

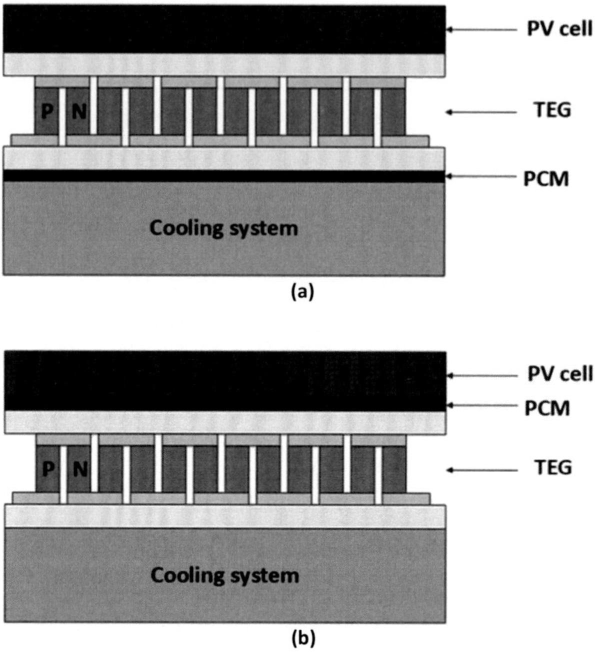

Figure 4.16 Two kinds of layouts of the PCM in a PV-TE system: (a) PV-TE-PCM system and (b) PV-PCM-TE system. Recreated based on [45, 46].

4.5.3 Novel Structure Design of a TEG in a PV-TE Hybrid System

Unlike the PV cell, the effect of geometric structure on a TEG's performance cannot be ignored. Normally, a TEG consists of three different parts: ceramic substrates, conductors, and cascade TE couples. By adjusting the layout of the three parts of a TEG, it is possible to improve the performance of a PV-TE system. Zhang and Xuan [47] verified the efficiency of a PV-TE system can be increased by simplifying the TEG structure in which the ceramic substrates are removed. Meanwhile, a V-type groove layout (as shown in Fig. 4.17) was proposed. Compared to the traditional and simplified PV-TE systems, the V-type groove layout can increase the efficiency of the hybrid system further.

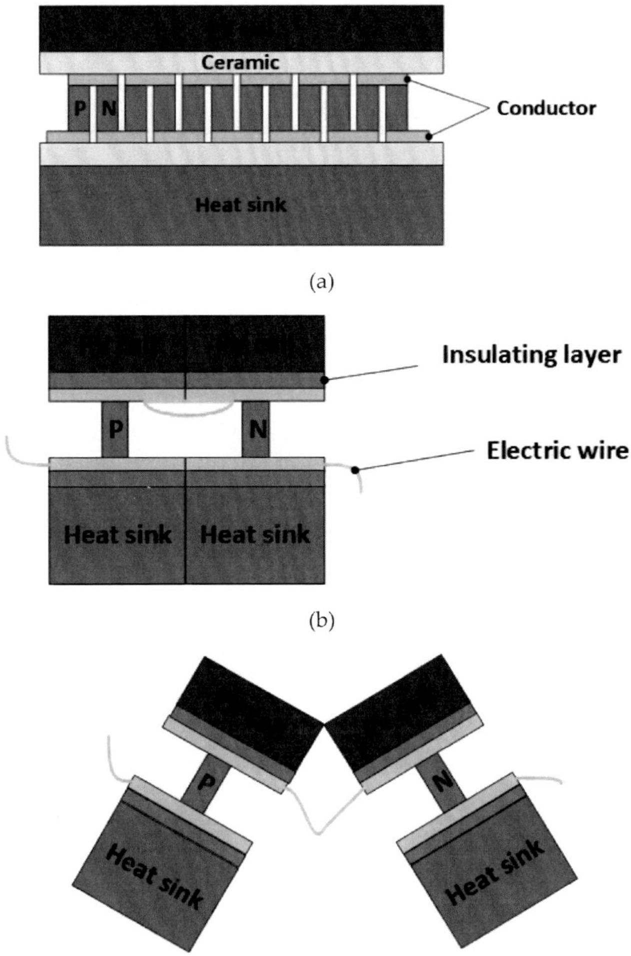

(a)

(b)

(c)

Figure 4.17 Three kinds of PV-TE hybrid systems: (a) traditional system, (b) simplified system, and (c) system with V-type groove. Recreated based on [47].

Moreover, previous studies have indicated that a TEG's power and efficiency can be increased by optimizing the shape of TE legs [51–53]. In this way, it is possible to apply an improved TEG module (such as a segment or variable cross-sectional TEG, as shown in Fig. 4.18) to increase the overall performance of a PV-TE hybrid system.

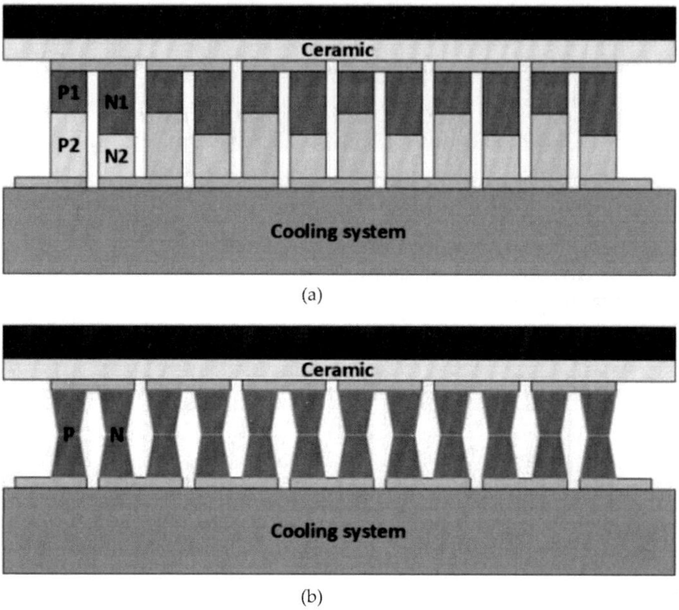

Figure 4.18 Two kinds of novel structure design of a TEG: (a) segment TEG module and (b) variable cross-sectional TEG module. Recreated based on [48, 49].

Cui et al. [48] utilized numerical methods to compare the performance of a segmented PV-TE system and the traditional PV-TE. The results indicate that there was an increase of 3.3% in the efficiency for the segmented PV-TE system with the concentrated solar radiance increased from 4 kW/m^2 to 24 kW/m^2. However, the efficiency in the hybrid systems with only PbTe and Bi_2Te_3 TE materials is increased by 1.2% and 1.7%, respectively, under the same situation. Besides, Maduabuchi [49] utilized the numerical method to analyze the performance of a PV-cross-sectional TE system. He found the variable cross-sectional TEG is more suitable for a PV-TE system than the constant cross-sectional one when the solar irradiance is insufficient.

4.6 Economic Analysis of a PV-TE System

The market competitiveness of an energy system is not only dependent on its performance in energy conversion but decided on cost-effectiveness. Currently, a large number of studies have reported that a PV-TE hybrid system has a better performance in power generation than that of a PV-only system under the same working conditions [17, 36]. However, compared to the PV-only device, a hybrid system consists of more components, such as TEG modules and splitters, which can lead to an obvious increase in the cost. Therefore, it is necessary to make an analysis of a PV-TE hybrid system from an economic aspect, which is beneficial to fully evaluate the prospect of this technology.

Based on a PV cell (total area is 1.2 × 0.18 m^2), Li et al. [54] proposed two kinds of hybrid schemes: integrated PV-TE system and micro-channel heat pipe (MCHP) PV-TE system, as shown in Fig. 4.19. An economic analysis was made by them to evaluate the prospects of the two hybrid schemes and a PV-only system. Firstly, considering an application scenario in Hefei (31°53' N and 117°15' E), a city in the eastern region of China, they calculated the annual electricity production from the PV-only system, integrated PV-TE system, and MCHP-PV-TE system, respectively. They found that the two hybrid systems produced more electricity than that of the PV-only system. This may be because the integrated PV-TE scheme was equipped with more TEG modules compared to the MCHP-PV-TE system. The results indicate that the annual electricity production of the integrated PV-TE system is almost two times as much as the MCHP-PV-TE system.

From the economic aspect, due to the greater number of TEG modules, the cost of the integrated PV-TE is much higher than that of the MCHP-PV-TE system. The extra cost of the components in the two hybrid systems can be found in Table 4.2. Through the cost calculation, the extra cost of the integrated PV-TE system is around 41 times higher than that of the MCHP-PV-TE system. In this way, considering the power generation and cost-effectiveness, the MCHP-PV-TE hybrid scheme has a better economic advantage compared to the integrated PV-TE system.

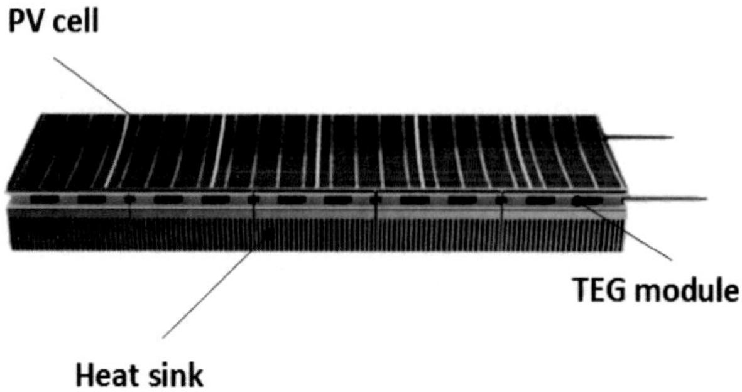

(a)

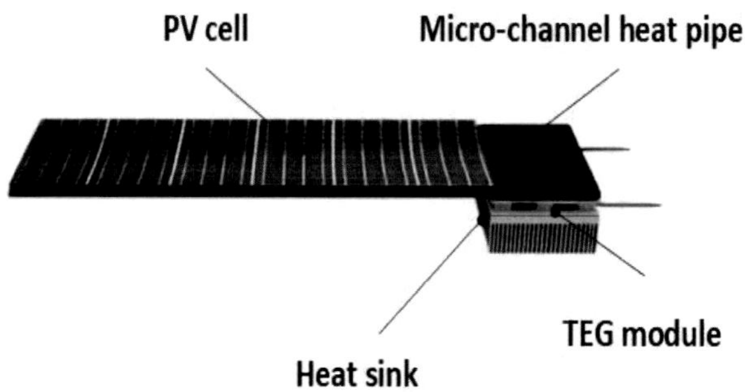

(b)

Figure 4.19 Structure diagram of the two PV-TE hybrid schemes: (a) integrated PV-TE system and (b) MCHP-PV-TE system. Recreated based on [54].

Table 4.2 Extra costs in the two PV-TE systems [54]

Hybrid system	Cost of MCHP	Cost of TEG	Cost of Heat Sink
PV-TE	–	￥22 (about \$3) × 71	￥1 (about \$0.14) × 71
MCHP-PV-TE	￥15 (about \$2)	￥22 (about \$3) × 1	￥1 (about \$0.14)

Moreover, in the economic life cycle of an energy system, the value of the total cash flows can be evaluated by the net present value (NPV), using Eq. (4.14) [54],

$$\mathrm{NPV} = \sum_{j=1}^{n} \frac{B_j - C_j}{(1+r)^j}, \tag{4.14}$$

in which, B and C are the future values of the benefit and cost, respectively, J is the time period, and the interest rate is symbolled as r.

Li et al. [54] utilized the NPV to compare the payback period (PBP) of the MCHP-PV-TE and PV-only systems. The PBP is actually the time period in years, calculated to make the NPV zero. Due to the use of the same type of PV cell in both MCHP-PV-TE and PV-only systems, they only evaluated the PBP of the extra investment for the MCHP system. According to their calculations, after six years, the extra investment can be recovered in the MCH-PV-TE system.

In addition to the increase in cost-effectiveness, the further development of hybrid PV-TE technology may have positive effects on the socio-economic aspect. As for emerging industries, exploiting renewable energy technologies is a useful way to increase social benefits, such as welfare and employment [55]. Referring to the trajectory of conventional solar technology, the generating cost of solar power experienced a decline of 79% during the period from 2010 to 2019, due to technical improvement (refer to Fig. 4.20). Meanwhile, solar power belongs to a labor-intensive industry, making it possible to create a large number of new job opportunities. The data in Fig. 4.21 indicates that there were about 3.6 million jobs related to the solar power industry in 2018. Compared to 2012, the number of employment positions increased almost 2.6 times.

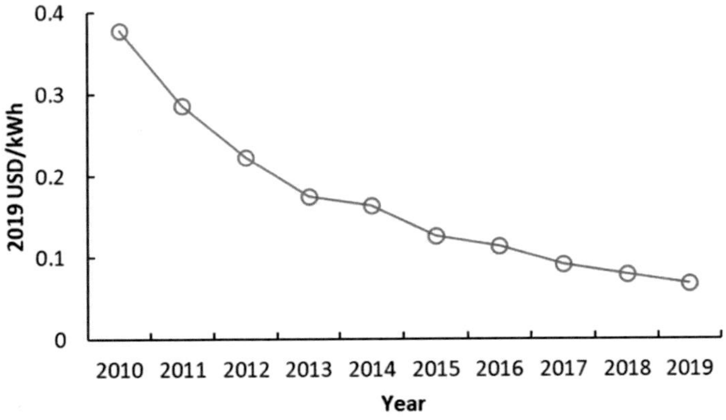

Figure 4.20 Variations of levelized cost effect of solar PV, 2010–2019. Recreated based on [56].

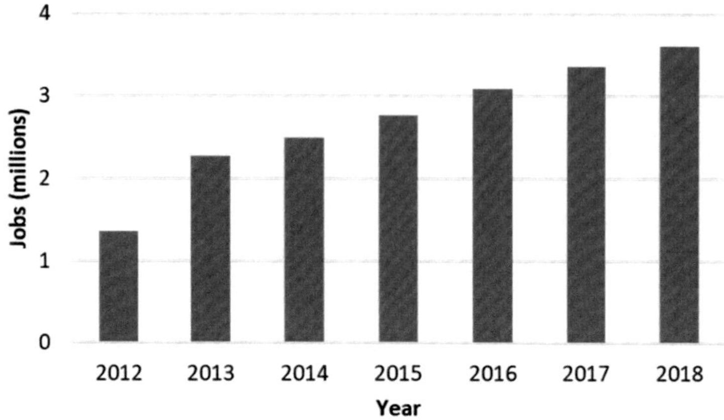

Figure 4.21 Job opportunities related to the solar PV industry, 2010–2019. Recreated based on [57].

4.7 Conclusions

It is acknowledged that improving efficiency is not only beneficial to decarbonization, but it can accelerate the commercialization process for renewable energy technology. As for a PV system, thermal recovery technology is considered as a useful way to improve its energy conversion efficiency. Equipped with a compact structure and zero emissions, the TEG is an excellent complement to recover the waste heat of a PV cell. This chapter commences with the working principles of a PV cell and TEG module, the core components of a PV-TE system. A hybrid PV-TE system can produce more power than a PV-only system, although adding TEGs may increase the temperature of PV cells slightly. The efficiency of a hybrid PV-TE system can be improved compared to that of a PV-only system, depending on the type of PV cell. For example, for a dye-sensitized PV cell, the efficiency even can be increased by 30% through hybrid TEG modules.

Recently, many novel design schemes for the PV-TE hybrid system were proposed to improve its performance. The main improvement schemes explore three aspects, which are optical design, heat transfer technology, and geometric structure. These studies indicate that there is much room for further improvement in the performance of the improved PV-TE hybrid system compared to the simple integrated hybrid scheme. Additionally, through economic analysis, the PV-TE hybrid system has been verified that it is possible to be a techno-economic solution for renewable energy. However, the economic advantage may differ among the specific hybrid schemes. Therefore, in future studies, it is necessary to consider the economic factors when improving the performance of a PV-TE hybrid system.

Acknowledgment

This work was made possible by funding from the Natural Sciences and Engineering Research Council of Canada.

References

1. International Energy Agency (IEA, 2022), Tracking Industry 2021, https://www.iea.org/reports/tracking-industry-2021. Accessed 23, October 2022.
2. International Energy Agency (IEA, 2022), Global Energy Review: CO_2 Emissions in 2021, https://www.iea.org/reports/global-energy-review-co2-emissions-in-2021-2. Accessed 23, October 2022.
3. International Energy Agency (IEA, 2022), Renewable Electricity, https://www.iea.org/reports/renewable-electricity. Accessed 23, October 2022.
4. Cheng, J. H., Yeh. C. H., and Tu C. W., (2008) Trust and knowledge sharing in green supply chains, *Supply Chain Management*, vol. 13, no. 4, pp. 283–295.
5. Zhai, P., Larsen, P., Millstein, D., Menon, S., and Masanet, E., (2012) The potential for avoided emissions from photovoltaic electricity in the United States, *Energy*, vol. 47, no. 1, pp. 443–450.
6. Tawalbeh, M., Al-Othman, A., Kafiah, F., Abdelsalam, E., and Almomani F., (2021) Environmental impacts of solar photovoltaic systems: A critical review of recent progress and future outlook, *Science of the Total Environment*, vol. 759, p. 143528.
7. Kalogirou, S. A. (2014) *Solar energy engineering: Processes and systems*, 2nd Ed. (Elsevier: Oxford, UK).
8. Emery, K., Burdick. J., Caiyem Y., and Dunlavy, D., et al., (1996) Temperature dependence of photovoltaic cells, modules, and

systems, in *Conference Record of the Twenty Fifth IEEE Photovoltaic Specialists Conference*, pp. 1275–1278.

9. Kaldellis, J. K., Kapsali, M., and Kavadias, K. A., (2014) Temperature and wind speed impact on the efficiency of PV installations. Experience obtained from outdoor measurements in Greece, *Renewable Energy*, vol. 66, pp. 612–624.
10. Rajkumar, V. A., Weijers, C., and Debije, M. G., (2015) Distribution of absorbed heat in luminescent solar concentrator lightguides and effect on temperatures of mounted photovoltaic cells, *Renewable Energy*, vol. 80, pp. 308–315.
11. Yang, B., Zeng, C., Li, D., and Guo, Z., et al., (2022) Improved immune genetic algorithm-based TEG system reconfiguration under non-uniform temperature distribution, *Applied Energy*, vol. 325, p. 119691.
12. Attivissimo, F., Di Nisio, A., Lanzolla, A. M. L., and Paul, M., (2015) Feasibility of a photovoltaic–thermoelectric generator: performance analysis and simulation results, *IEEE Transactions on Instrumentation and Measurement*, vol. 64, no. 5, pp. 1158–1169.
13. Zhang, J., Liu, Y., and Yang, L., (2014) Performance estimation of photovoltaic–thermoelectric hybrid systems, *Energy*, vol. 78, pp. 895–903.
14. Pang, W., Liu, Y., and Gao, X., (2015) Empirical study on thermal performance through separating impacts from a hybrid PV/TE system design integrating heat sink, *International Communications in Heat and Mass Transfer*, vol. 60, pp. 9–12.
15. Wehbi, W., Taher, R., Faraj, J., and Castelain, C., (2022) Hybrid thermoelectric generators-renewable energy systems: A short review on recent developments, *Energy Reports*, vol. 8, pp. 1361–1370.
16. Van Sark, W. G. J. H. M. (2011) Feasibility of photovoltaic – Thermoelectric hybrid modules, *Applied Energy*, vol. 88, pp. 2785–2790.
17. Li, G., Shittu, S., Diallo, T. M. O., and Yu, M., et al., (2018) A review of solar photovoltaic-thermoelectric hybrid system for electricity generation, *Energy*, vol. 158, pp. 41–58.
18. Kandil. A. A., Awad, M. M., Sultan, G. I., and Salem, M. S., (2022) Investigating the performance characteristics of low concentrated photovoltaic systems utilizing a beam splitting device under variable cutoff wavelengths, *Renewable Energy*, vol. 196, pp. 375–389.

19. Li, Y., (2022) A concentrated solar spectrum splitting photovoltaic cell-thermoelectric refrigerators combined system: Definition, combined system properties and performance evaluation, *Energy*, vol. 238, p. 122042.

20. He, H., Zhao, Y., Ren, H., and Niu, C., et al., (2022) Novel principle for characterizing the material-level parameters of a thermoelectric generator module, *Applied Thermal Engineering*, vol. 215, p. 118900.

21. Lan, Y., Lu. J., Li, J., and Wang, S., (2022) Effects of temperature-dependent thermal properties and the side leg heat dissipation on the performance of the thermoelectric generator, *Energy*, vol. 243, p. 123035.

22. Olivares-Robles, M. A., Badillo-Ruiz, C. A., and Ruiz-Ortega, P. E., (2020) A comprehensive analysis on nanostructured materials in a thermoelectric micro-system based on geometric shape, segmentation structure and load resistance, *Scientific Reports*, vol. 10, no. 1, p. 21659.

23. Doraghi, Q., Khordehgah, N., Żabnieńska-Góra, A., and Ahmad, L., et al., (2021) Investigation and computational modelling of variable TEG leg geometries, *ChemEngineering*, vol. 5, no. 3, p. 45.

24. Babu, C. and Ponnambalam, P., (2018) The theoretical performance evaluation of hybrid PV-TEG system, *Energy Conversion and Management*, vol. 173, pp. 450–460.

25. Rajaee, F., Rad, M. A.V., Kasaeian, A., and Mahian, O., et al., (2020) Experimental analysis of a photovoltaic/thermoelectric generator using cobalt oxide nanofluid and phase change material heat sink, *Energy Conversion and Management*, vol. 212, p. 112780.

26. Yang, B., Wang, J., Zhang, X., and Yu, T., et al., (2020) Comprehensive overview of meta-heuristic algorithm applications on PV cell parameter identification, *Energy Conversion and Management*, vol. 208, p. 112595.

27. Lamba, R. and Kaushik, S. C., (2016) Modeling and performance analysis of a concentrated photovoltaic–thermoelectric hybrid power generation system, *Energy Conversion and Management*, vol. 115, pp. 288–298.

28. Nagayoshi, H., Orio, S., Kono, Y., and Nakajima, H., (2002) Novel PV array/module *I–V* curve simulator circuit, in *Conference Record of the Twenty-Ninth IEEE Photovoltaic Specialists Conference*, pp. 1535–1538.

29. He, W., Zhang, G., Zhang, X., and Ji, J., et al., (2015) Recent development and application of thermoelectric generator and cooler, *Applied Energy*, vol. 143, pp. 1–25.

30. Jouhara, H., Żabnieńska-Góra, A., Khordehgah, N., and Doraghi, Q., et al., (2021) Thermoelectric generator (TEG) technologies and applications, *International Journal of Thermofluids*, vol. 9, p. 100063.

31. Karri, M. A., Thacher, E. F., and Helenbrook, B. T., (2011) Exhaust energy conversion by thermoelectric generator: Two case studies, *Energy Conversion and Management*, vol. 52, no. 3, pp. 1596–1611.

32. Chen, L., Li, J., Sun, F., and Wu, C., (2005) Performance optimization of a two-stage semiconductor thermoelectric-generator, *Applied Energy*, vol. 82, no. 4, pp. 300–312.

33. Maduabuchi, C., Eneh, C., Alrobaian, A. A., and Alkhedher, M., (2023) Deep neural networks for quick and precise geometry optimization of segmented thermoelectric generators, *Energy*, vol. 263, p. 125889.

34. Ge, M., Zhao, Y., Li, Y., and He, W., et al., (2022) Structural optimization of thermoelectric modules in a concentration photovoltaic–thermoelectric hybrid system, *Energy*, vol. 244, p. 123202.

35. Singh, S., Ibeagwu, O. I., and Lamba, R., (2018) Thermodynamic evaluation of irreversibility and optimum performance of a concentrated PV-TEG cogenerated hybrid system, *Solar Energy*, vol. 170, pp. 896–905.

36. Liwanag, M., Montayre, M. L. O., and Macabebe, E. Q. B., (2015) Feasibility study on thermoelectric conversion to improve photovoltaic operation, in *Conference Solar World Congress*, pp. 1–11.

37. Khan, M., Khan, M. I., Kazim, A. H., and Shabir, A., et al., (2021) An experimental and comparative performance evaluation of a hybrid photovoltaic-thermoelectric system, *Frontiers in Energy Research*, vol. 9, p. 722514.

38. Shou, C., Luo, Z., Wang, T., and Shen, W., et al., (2012) Investigation of a broadband TiO_2/SiO_2 optical thin-film filter for hybrid solar power systems, *Applied Energy*, vol. 92, pp. 298–306.

39. Mahmoudinezhad, S., Cotfas, D. T., Cotfas, P. A., and Skjølstrup, E. J. H., et al., (2022) Experimental investigation on spectrum beam splitting photovoltaic–thermoelectric generator under moderate solar concentrations, *Energy*, vol. 238, p. 121988.

40. Alnajideen, M. and Min, G., (2022) Hybrid photovoltaic-thermoelectric system using a novel spectral splitting solar concentrator, *Energy Conversion and Management*, vol. 251, p. 114981.

41. Zhang, J., Zhai, H., Wu, Z., and Wang, Y., et al., (2020) Enhanced performance of photovoltaic–thermoelectric coupling devices with thermal interface materials, *Energy Reports*, vol. 6, pp. 116–122.
42. Makki, A., Omer, S., Su, Y., and Sabir, H., (2016) Numerical investigation of heat pipe-based photovoltaic–thermoelectric generator (HP-PV/TEG) hybrid system, *Energy Conversion and Management*, vol. 112, pp. 274–287.
43. Shittu, S., Li, G., Zhao, X., and Akhlaghi, Y. G., et al., (2019) Comparative study of a concentrated photovoltaic-thermoelectric system with and without flat plate heat pipe, *Energy Conversion and Management*, vol. 193, pp. 1–14.
44. Gao, Y., Chen, B., Wu, D., and Dai, Z., et al., (2022) Comparative study of various solar power generation systems integrated with nanofluid-flat heat pipe, *Applied Energy*, vol. 327, p. 120052.
45. Motiei, P., Yaghoubi, M., and GoshtasbiRad, E., (2019) Transient simulation of a hybrid photovoltaic-thermoelectric system using a phase change material, *Sustainable Energy Technologies and Assessments*, vol. 34, pp. 200–213.
46. Luo, Z., Zhu, N., Hu, P., and Lei, F., et al., (2022) Simulation study on performance of PV-PCM-TE system for year-round analysis, *Renewable Energy*, vol. 195, pp. 263–273.
47. Zhang, J. and Xuan, Y., (2019) An integrated design of the photovoltaic-thermoelectric hybrid system, *Solar Energy*, vol. 177, pp. 293–298.
48. Cui, Y. J., Wang, B. L., Li, J. E., and Wang, K. F., (2020) Performance evaluation and lifetime prediction of a segmented photovoltaic-thermoelectric hybrid system, *Energy Conversion and Management*, vol. 211, p. 112744.
49. Maduabuchi, C., (2022) Improving the performance of a solar thermoelectric generator using nano-enhanced variable area pins, *Applied Thermal Engineering*, vol. 206, p. 118086.
50. Cotfas, D. T., Cotfas, P. A., Mahmoudinezhad, S., and Louzazni, M., (2022) Critical factors and parameters for hybrid photovoltaic-thermoelectric systems: A review, *Applied Thermal Engineering*, vol. 215, p. 118977.
51. Ming, T., Wu, Y., Peng, C., and Tao, Y., (2015) Thermal analysis on a segmented thermoelectric generator, *Energy*, vol. 80, pp. 388–399.
52. Ge, Y., He, K., Xiao, L., and Yuan, W., et al., (2022) Geometric optimization for the thermoelectric generator with variable cross-section legs by coupling finite element method and optimization algorithm, *Renewable Energy*, vol. 183, pp. 294–303.

53. Liu, H. B., Wang, S. L., Yang, Y. R., and Chen, W. H., et al., (2020) Theoretical analysis of performance of variable cross-section thermoelectric generators: Effects of shape factor and thermal boundary conditions, *Energy*, vol. 201, p. 117660.

54. Li, G., Zhao, X., and Ji, J., (2016) Conceptual development of a novel photovoltaic-thermoelectric system and preliminary economic analysis, *Energy Conversion and Management*, vol. 126, pp. 935–943.

55. Seker, S. and Kahraman, C., (2021) Socio-economic evaluation model for sustainable solar PV panels using a novel integrated MCDM methodology: A case in Turkey, *Socio-Economic Planning Sciences*, vol. 77, p. 100998.

56. International Renewable Energy Agency (IRENA, 2023), Renewable Power Generation Costs in 2019, https://www.irena.org/publications/2020/Jun/Renewable-Power-Costs-in-2019. Accessed 25, April 2023.

57. Renewable Energy Agency (IRENA, 2023), Renewable Energy and Jobs Annual Review 2019, https://www.irena.org/publications/2019/Jun/Renewable-Energy-and-Jobs-Annual-Review-2019. Accessed 25, April 2023.

Chapter 5

Barriers to the Deployment of Renewable Energy Technologies

Shahla Choobchian,[a] Mahnaz Ghorbannejad,[a] Mohammad Shokati Amghani,[a] and Hossein Azadi[b]

[a]*Department of Agricultural Extension and Education, College of Agriculture, Tarbiat Modares University (TMU), Tehran 1497713111, Iran*
[b]*Department of Economics and Rural Development, Gembloux Agro-Bio Tech, University of Liège, Belgium*

shchoobchian@modares.ac.ir

Iran, like many other countries, is faced with the critical problems of limited fossil fuel resources and the negative effects of greenhouse gas emissions. These challenges have become more critical due to Iran's expulsion from the ranks of oil-exporting countries in the current century, which has disrupted conventional financial sources and impacted livelihoods. In this context, renewable energies emerge as a viable solution to address various challenges in Iran. The rapid increase in energy demand, the limitation of fossil energy resources, and concerns about the economic and environmental consequences have required

Reclaiming Eden: Responsible Living, Engineering, and Architectures
Edited by David S.-K. Ting and Jacqueline A. Stagner

ISBN 978-981-5129-31-1 (Hardcover), 978-1-003-49612-0 (eBook)
www.jennystanford.com

countries to use renewable energy technologies. In Iran also, renewable energy technologies can play an essential role in reducing greenhouse gas (GHG) emissions, providing energy security, and supplying a solution to many problems. Therefore, the present study aims to identify developmental barriers to renewable energy technologies in Iran. In this regard, a literature review and face-to-face interviews are carried out to identify barriers. As a result, 28 barriers were identified. Four institutional, social, economic, and infrastructural dimensions have been considered to group the identified factors. Forty-four specialists in renewable energy technologies are interviewed to rank the identified barriers in the order of priority. Findings show that the main factor in preventing renewable energy penetration is a political and regulatory framework, with a focus on fossil fuels. Other effective factors are high initial investment costs, poor awareness of people, and lack of local infrastructure. Accordingly, ranking barriers in order of importance provides a mechanism for Iranian decision-makers and policymakers to develop policies in line with accepting renewable energy.

5.1 Introduction

In developing countries, non-renewable energy (RE) sources, such as oil and gas, contribute to around 90% of net energy resources, making them dependent on these energy resources (International Energy Agency, 2011; Afful-Dadzie et al., 2020) and causing severe environmental global problems (Sansaniwal et al., 2017; Yaghoubi et al., 2019). The process of the depletion of fossil fuel resources and oil reserves is very fast, being the major cause of harmful gas emissions (Dyer & Desjardins, 2018). Some of the negative effects of these gases are the receding of glaciers, and loss of biodiversity, climate change, sea level increase, etc. (Sansaniwal et al., 2017; Yazdanpanah et al., 2022). Another effect of increased fossil fuel consumption is its adverse effect on global economic activities, due to the increase a crude oil price. Industrialization and motorization are two important driving factors of the modern era leading to unpredictable fuel demand (Agarwal, 2007). According to energy use estimations around the world, carbon dioxide (CO_2) emissions will propose an annual increase of 1.7% by 2030, reaching about 38 billion tons (World Energy Outlook, 2017; Li & Wang, 2017). One of the proposed strategies for limiting climate change is greenhouse emission stabilization (United Nations Development Program, 2007; Shokri, 2012). Developing a low-carbon economy and formulating a policy for carbon dioxide emission reduction is of great importance for policymakers and researchers in the present energy trends (International Energy Agency, 2015). Intensive use of carbon-based fossil fuels has had adverse environmental effects that have made many countries develop different programs to change the exploitation system of non-renewable to renewable energies (Akbi et al., 2017; Rosen, 2009). Renewable energies have a significant potential for growth. This is particularly the case in scenarios where there are environmental constraints. Some of these constraints are CO_2 emissions and global warming. Despite having the potential for RE, Iran, as a developing country, is highly dependent on conventional energy sources (about 98% of the net energy resources) (International Energy Agency, 2011; Dehhaghi et al., 2022). Adverse effects of non-renewable energy utilization are soil infertility, increased greenhouse gas effects, and deforestation (United Nations Environment Programme, 2015). Population growth has increased energy demand and non-renewable energy resources do not suffice to meet this need (Intergovernmental Panel on Climate Change, 2014; Laha & Chakraborty, 2017; Momenpour & Choobchian, 2022). At the same time, RE has a small share of Iran's energy portfolio (Fadai et al., 2011; Dehhaghi et al., 2022). Therefore, it is vital to identify major obstacles to clean energy development in Iran.

The limited fossil fuels and the problems associated with greenhouse gas emissions have made clear the need for more and more attention to renewable energies (Dehhaghi & Choobchian, 2020). The high growth of energy consumption in Iran and the exodus of major oil and gas exporters since the end of this century seriously affect the country's development. Additionally, short-term planning based on the current economic decision-making model can be the biggest obstacle to the growth of RE production in many developing countries. This research endeavor aims to assist Iran in attaining its objectives of enhancing energy security and reducing CO_2 emissions by identifying the barriers, phasing out the utilization of fossil fuels, and substituting them with renewable energy sources. In this regard, the main objective of this paper is twofold. First, it aimed to identify obstacles to the development of clean energy technologies in Iran. Second, it aimed to rank identified obstacles to the development of clean energy technologies. According to these facts, developing countries like Iran heavily depend on non-renewable energy sources, leading to environmental and economic challenges (Noorani et al., 2022). Therefore, the rapid depletion of fossil fuel reserves, greenhouse gas emissions, and increased energy demand highlight the need for transitioning to RE. Despite its potential, Iran relies mainly on conventional energy sources. This research seeks to identify and prioritize obstacles to clean energy technology development in Iran, aiming to facilitate the transition toward sustainable, low-carbon energy solutions.

5.2 Research Background

Within a study titled "Analyzing the Causes of Non-Development of RE-Related Industries in Iran," factors such as the lack of needed funding for the implementation of projects, the lack of specialized technical consultants, and spending too much time on technological deployment have been identified as major obstacles (Fadai et al., 2011).

In another research entitled "A Review of Current Energy Systems and Green Energy Potential in Kazakhstan," it was found that the government focuses on fossil fuels, investment issues, and immature legal frameworks to stimulate clean energy use in the electricity sector. In addition, it was found that insufficient environmental knowledge and economic profits from clean energies prevent the widespread diffusion of related technologies (Karatayev & Clarke, 2016).

Sindhu et al. (2016) carried out a study entitled "Identification and Analysis of Barriers to Implementation of Solar Energy in Indian Rural Sector Using Integrated ISM and Fuzzy MICMAC Approach" and found that high initial financial costs, low productivity (efficiency) of installation projects, credit, and security are the main barriers. The removal of these high-level barriers depends on removing other barriers, including legal and institutional barriers.

Luthra et al. (2015) in their study titled "Barriers to Renewable/Sustainable Energy Technologies Adoption: Indian Perspective," ranked experts' opinions according to the hierarchical analysis method. They found that high costs of investment in economic aspects, lack of sufficient market in market dimension, and technologically, the complexity of these technologies are important obstacles. In addition, the lack of subsidies is a serious obstacle in the economic dimension.

Richards et al. (2012) conducted a study entitled "Barriers to Wind Energy Development in Saskatchewan (Canada)" and found that from the viewpoint of stakeholders, political and technological barriers are the major obstacles. This is while perceived technological and political barriers were due to a lack of knowledge or explicit satisfaction with the status quo.

In another survey study, key obstacles to the placement of solar energy systems in Hong Kong were identified as economic, infrastructural, social, and regulatory (Zhang et al., 2012).

Abas et al. (2021) highlighted the challenges and economic considerations associated with integrating RE sources into Iran's energy infrastructure. They emphasized the need for grid enhancements and cost-effective strategies to facilitate this transition. In addition, Zare et al. (2020) studied about the deployment of solar photovoltaic technology in Iran. They found that solar PV projects demonstrated economic feasibility and contributed to reducing greenhouse gas emissions, serving as a promising clean energy solution. Furthermore, Eslamian et al. (2018) conducted a geographical assessment of geothermal energy potential across Iran. They identified regions with significant geothermal resources, highlighting opportunities for sustainable heating and electricity generation. In another study, Mahdavi and Yarahmadi (2014) assessed the status of wind energy in Iran. They reported substantial wind energy potential, particularly in certain regions, and discussed technological advancements that contribute to its growth.

The dominance of conventional fuels in Iran and therefore increasing concerns about its side effects (both environmental issues and economic) has required the government and policymakers to switch to the adoption of renewable technologies.

However, facing some obstacles has lowered the adoption rate of these technologies (Oryani et al., 2021). From the surveyed literature, the importance of RE depends on the socio-economic, technological, and technical potential, geographic (location), and political circumstances. Consequently, the lack of studying these impediments may lead to the formulation of inappropriate related policies. Therefore, understanding and identifying the barriers is the necessity of designing suitable strategies.

Despite the enormous benefits of renewable energies, their development and utilization face many obstacles. Such barriers in a country may be similar to those in other countries, both developed and developing ones (Karatayev et al., 2016). In this regard, Zhang et al. (2012) and Reddy & Painuly (2004) described the general characteristics of clean energy challenges in the world. While several studies have investigated the impediments to the adoption of energy-efficient technologies in different developing countries, little research has been conducted on Iran. Therefore, the result of this research may provide the basic framework for the Government, the policymakers, the manager, the private sector, and the executors involved in designing and formulating inclusive RE projects in developing countries like Iran. A comprehensive literature review on the barriers to the development of RE has been presented in this study, which can assist interested researchers in this field. The most crucial contribution of this paper could be attributed to identifying and ranking the hindrances to the development of RE in Iran to overcome the barriers and find a way to increase its share.

It is worth noting that there have been no studies ranking barriers to RE technology deployment. In addition, there has been no study that has identified and ranked items of barriers separately; this prioritization is essential for planning and policymaking at the macro and micro levels. According to what was said, this article is going to answer these questions:

(1) What is the most essential obstacle to clean energy penetration in developing countries such as Iran?

(2) How are obstacles prioritized?

In conclusion, this literature review underscores the critical importance of transitioning toward RE sources within the context of Iran, where the challenges of limited fossil energy resources and the adverse effects of greenhouse gas emissions demand a transformative response. The urgency of this transition has increased as a result of Iran's expulsion from oil-exporting countries in the 21st century because the country deals with the consequences on both a social and economic basis. This comprehensive review compiled findings from a variety of research, each of which provided insight into different aspects of Iran's clean energy landscape, ranging from solar and geothermal potential to economic feasibility and political frameworks. While global trends emphasize the imperative of renewable energy, this study contributes uniquely by identifying and categorizing 28 developmental barriers specific to Iran's context, and subsequently, ranking these barriers by priority. The greatest obstacle highlighted, the political and legal structure based on fossil fuel dependency, emphasizes the importance of specific measures that promote renewable energy adoption. This case study is an important resource for Iranian policymakers and decision-makers, which is made possible by combining these insights. It provides an approach for overcoming the complex difficulties associated with sustainable energy development while addressing urgent environmental and energy security concerns.

According to the background of the research, the most important variables of the present study can be classified and analyzed as RE development barriers. The barriers include economic, infrastructural, institutional, and social barriers (Fig. 5.1).

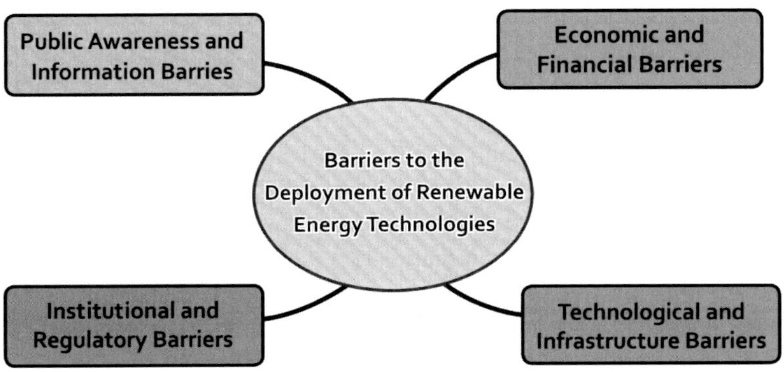

Figure 5.1 Theoretical framework.

5.3 Status of Energy in Iran

Iran was the fifth-largest crude oil producer in OPEC in 2021 and the third-largest natural gas producer in the world in 2020. It holds some of the world's largest deposits of proved oil and natural gas reserves, ranking as the world's third-largest oil and second-largest natural gas reserve holder in 2021. At the end of 2021, Iran accounted for 24% of oil reserves in the Middle East and 12% in the world. Despite its abundant reserves, Iran's crude oil production has fallen since 2017 because the oil sector has been subject to underinvestment and international sanctions for several years (Zahedi et al., 2022).

Although Iran is a member of OPEC, it is exempt from the production cuts under the OPEC+ agreement because its crude oil production is constrained as a result of sanctions. Iran's crude oil production reached a 30-year low in 2020, as a result of these sanctions and the economic impacts of the global COVID-19 pandemic. Output rose slightly in 2021 because global oil demand increased. Although sanctions on its oil exports remained in place, Iran shipped more crude oil, primarily to China, in 2021. If sanctions were lifted, Iran's crude oil production could return to full capacity, which EIA assesses at 3.7 million barrels per day (b/d). Indirect negotiations related to Iran's nuclear program between the United States and Iran began in April 2021 and are ongoing as of September 2022 (Ghouchani et al., 2021).

Iran's economy is relatively diversified compared with many other Middle Eastern countries, but petroleum and other liquids exports are a significant source of government revenue in 2021, Iran's oil companies earned about $40 billion in net oil export revenues, up from around $15 billion in 2020. Total export revenues increased in 2021 as a result of rising global oil prices and Iran's total petroleum liquids exports increasing from 2020. We estimate the oil price increases in 2022 will further increase Ira's revenues (Jahangiri et al., 2019).

Iran's economy consumed an estimated 11.6 quadrillion British thermal units of primary energy in 2021, making it the highest energy consumer in the Middle East. Natural gas and oil accounted for almost all of Iran's total primary energy consumption, and hydropower, coal, nuclear, and non-hydropower renewables accounted for the remaining shares (Zahedi et al., 2021).

5.3.1 Fossil Fuel Resources and CO_2 Emission in Iran

Iran has the largest resources of fossil fuels, such that it is ranked as the fourth oil-rich country and the second holder of natural gas reserves in the world. (Cheraghi et al., 2020). Additionally, Iran is ranked among the top 10 and 5 oil and gas producers in the world, respectively (EIA, 2015) (Figs. 5.2 and 5.3). About 10% of crude oil reserves around the world are held in Iran.

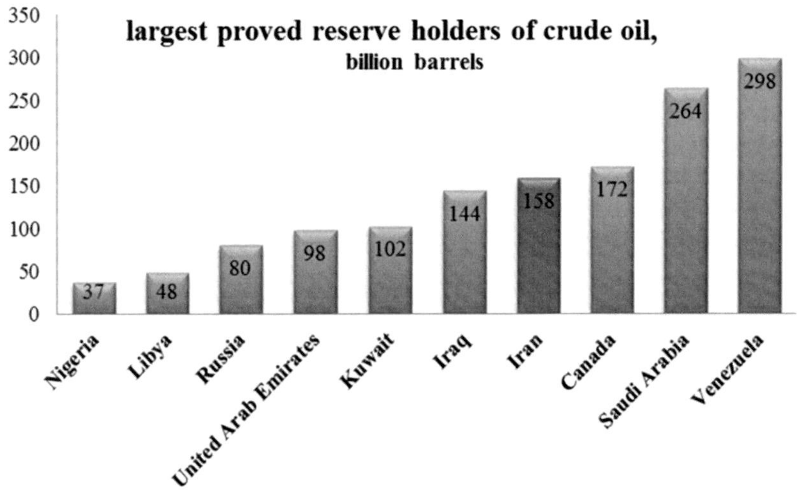

Figure 5.2 Top 10 crude oil reserves in the world (unit: billion barrels).

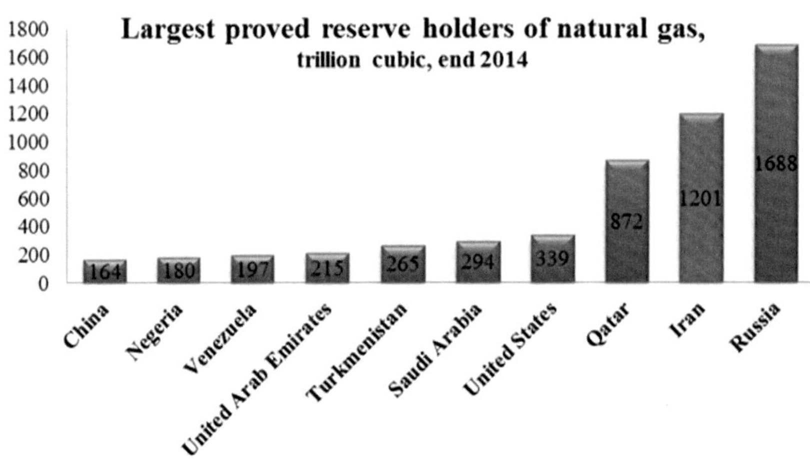

Figure 5.3 Top 10 natural gas holders (unit: trillion cubic).

In Iran, more than 70% of crude oil resources are placed onshore, and the rest are situated within the Gulf. Throughout the world, Iran is ranked second in terms of holding diversified gas reserves. Iran has approximately 17% of natural gas reserves across the world and one-third of OPEC's reserves. Nearly 40% of Iran's gas reserves are located in South Pars, the largest natural gas field in Iran (EIA, 2015) (Fig. 5.3).

In 2016, Iran consumed over 270 million tons of oil (EIA, 2018). Most of the primary energy consumption in Iran (about 98%) that has had a quick growth over the past decades belongs to gas and oil. Other marginal fuels are coal, nuclear, hydropower, and non-hydro renewable energies. Increased energy consumption has been despite the depression of economic growth and the imposition of nuclear-related sanctions. Primary energy consumption in Iran had an increase of about 40% between 2006 and 2016. Energy subsidy reform in Iran was developed to properly manage increased domestic energy demand growth and reduce the fund exposure to high subsidy prices, the result of which was increased domestic crude oil, gas, and electricity prices between 2010 and 2014. Iran's per capita energy consumption is more than the worldwide average consumption, i.e., 14 and 5 times more than Japan and India, respectively (Energy Balances, 2011). Iran is ranked 13 regarding energy consumption (Aslani et al., 2012). It is estimated that electricity demand will increase highly in 2030, which cannot be met by fossil fuels (Ghorashi & Rahimi, 2011). Increased energy demand in the future can be seen by adopting some new approaches by the Iranian government for the use of renewable energies. Electricity production from different sources in Iran in 2017 has been depicted in Fig. 5.4.

Natural gas plays an important role in electricity production from different sources in Iran. The share of gas is 71%. Meanwhile, the share of oil is 22%; hydropower, 5%; nuclear energy, 2%; and coal, wind, and biomass, 0%. Figure 5.5 shows the significant role of natural gas in electricity production in Iran. Figure 5.6 shows that in 2013, a quarter of the total energy has been consumed by the transportation sector in Iran. It also shows that this sector is placed after the residential, commercial, and public sectors

(34%) and is equal to industry (26%). In addition, non-energy uses (10%), agriculture (4%), and other sectors (0%) are placed next. Characteristics of having a resource-based economy and the structure of production based on energy consumption caused the consumption of fossil fuels at a high level, which is still on the rise.

CO_2 emissions in Iran have increased along with its growing economy since 1965, as a result of industrialization, increased natural gas end-users, and road transportation development. This is also because of the lack of improvement in energy efficiency in the economic sector.

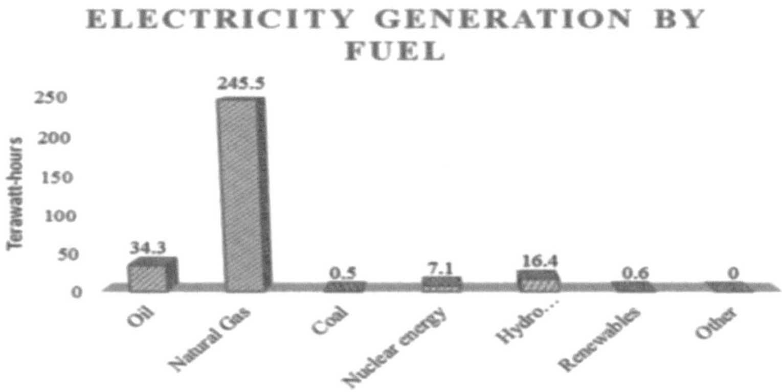

Figure 5.4 Electricity production in Iran, 2017 (unit: terawatt-hours).

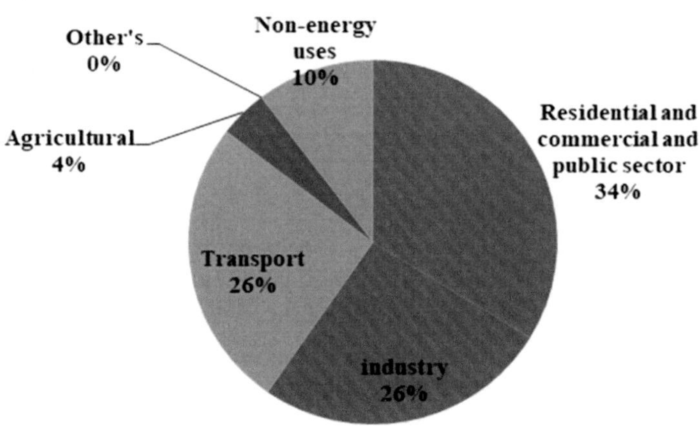

Figure 5.5 Electricity generation from different sources in Iran (International Energy Agency, 2014).

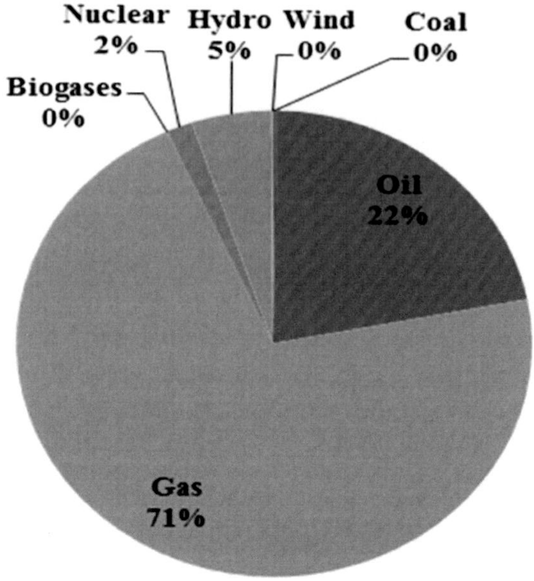

Figure 5.6 Final total energy consumption by different sectors by sector, 2013 (Energy Balance Sheet, 2014) (unit: percent).

Greenhouse gas emissions are generally due to fossil fuel consumption (Cheraghi et al., 2019). The Kyoto Protocol (KP) limitations are not observed in Iran due to high CO_2 emissions from different resources. That is why, Iran is ranked 9th in terms of generating carbon dioxide throughout the world (Hosseini et al., 2013). According to the IEA, Iran accounted for nearly two-thirds of global carbon dioxide emissions in 2015. In addition, per capita CO_2 emissions in Iran were 7, 11, 13, and 14 times higher than the United States, Germany and Finland, Italy, and Austria, respectively (EEA, 2015) (Fig. 5.7). According to the 2015 global CO_2 emission report, the change in per capita CO_2 emissions from 1990 to 2014 in Iran was 39% (Trends in Global CO_2 Emissions, 2015). The high intensity of CO_2 emission is first because of heavy industries, the high energy intensity of GDP, and the non-efficient use of energy. Second, the dominance of gas over oil is the reason for the intensity of carbon in the energy supplied. Using fossil fuels to provide energy in the transportation sector is the main reason for CO_2 emissions.

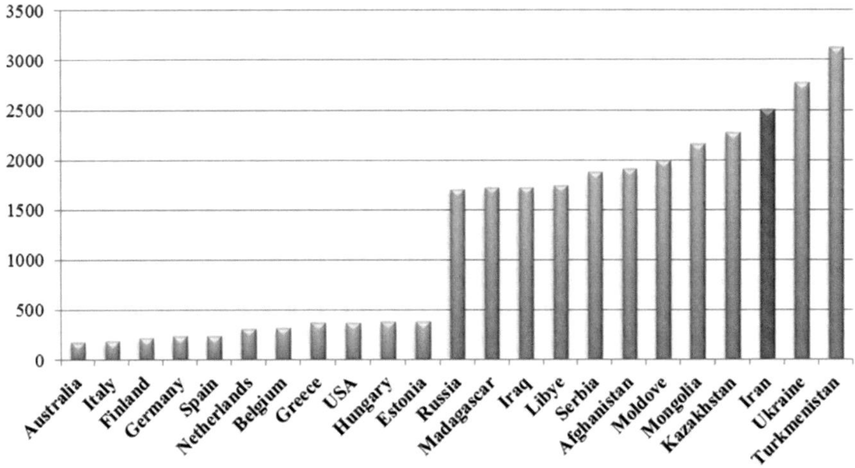

Figure 5.7 CO_2 emissions per dollar of GDP (European Environment Agency, 2015).

Due to the high proportion of fossil fuel and fossil energy consumption in Iran, and consequently, greenhouse gas emissions, all experts have suggested utilizing renewable energies rather than fossil energies. Accordingly, it was attempted to identify their barriers in Iran.

5.3.2 Potential for Clean Energy in Iran

5.3.2.1 Solar energy

Iran benefits from a significant solar energy potential, owing to its geographical location within the solar belt of the Earth. This region receives some of the highest levels of solar radiation globally, making it an ideal location for harnessing solar energy. According to data from Najafi et al. (2015), Iran has an abundance of solar radiation, with yearly solar radiation hours exceeding 2800 hours in many parts of the country.

The availability of abundant solar resources leads Iran to invest in solar energy for various applications, including electricity generation through photovoltaic systems and solar thermal technologies. This vast potential aligns with global efforts to transition toward sustainable energy sources and reduce reliance on fossil fuels (Mohammadi & Khorasanizadeh, 2019).

5.3.2.2 Wind energy

There are powerful air flows in several winter and summer months in Iran because of its geographical conditions and low air pressures. In winter, air pressures in the Atlantic Ocean and the Mediterranean Sea vary and result in cold northerly winds. In addition, eastern winds blowing from the Indian Ocean and western winds blowing from the Atlantic Ocean are the most significant reasons for 120 days of wind in Sistan. Initial estimations of wind energy in Iran show a capability of about 6.5 GW. However, there are studies suggesting the increase of up to 15 GW of wind potential in Iran (Najafi & Ghobadian, 2011). Iran is the only country within this geographic region with large-scale wind turbines.

5.3.2.3 Geothermal energy

In 1975, there were attempts for the exploration of geothermal resources in Iran. Substantial geothermal potential has been observed in the northern provinces of Iran. In addition, Iran has many hot water springs. Some of them have a temperature of about 85 °C. Investigations show that the temperature of deep wells reaches 150 °C. In Iran, 8.8% of sites are appropriate for extracting geothermal energy (Najafi and Ghobadian, 2011).

5.3.2.4 Biomass energy

Iran's biomass energy potential is estimated to be around 3500 MW by 2050 (SATBA, 2021). Although generating electricity from waste is an expensive method, the government guarantees to purchase the electricity produced by these power plants at a reasonable price to encourage investment in this field. In Iran, due to the variety of climate, more than 80% of agricultural products are agronomic products and the rest are horticultural products. The main agricultural products of Iran include wheat, corn, sugarcane, beets, and lucerne from 16.5 million hectares of cultivated land (Ministry of Agriculture Jihad, 2019). The amount of waste produced from agricultural products would be about 24 million tons per year (Solaymani, 2021) and would be a huge source for use in CHP plants.

Figure 5.8 shows the share of renewable energy capacity and electrical energy efficiency until 2020.

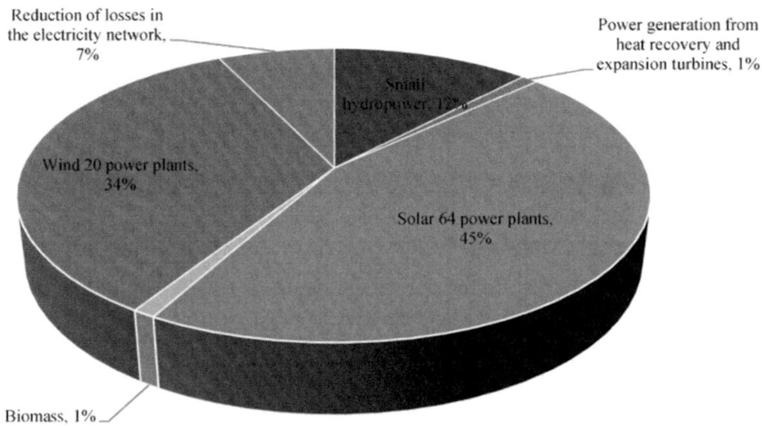

Figure 5.8 Share of renewable energy capacity and electrical energy efficiency in Iran (Azadi et al., 2017).

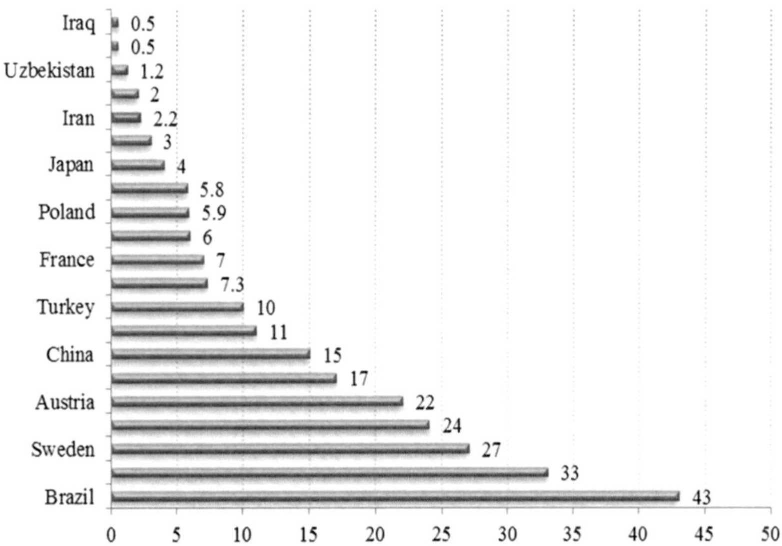

Figure 5.9 Renewable energy contribution to TPES.

Despite the potential for the production of different kinds of clean energies in Iran, their contribution to the energy mix of Iran is too low and accounts for a small part of energy production (Dehhaghi et al., 2022). Meanwhile, only 2.2% of TPES in Iran belongs to RE sources (Fig. 5.9). Iran can produce greater amounts of energy using renewable resources in comparison to several OECD. This is while the potential is not actualized and the involvement of clean energy in the energy basket of this country is about 2%.

5.4 Materials and Methods

5.4.1 Data Collection

Interviews were conducted with 44 experts in the area of clean energy technologies to prioritize the obstacles identified. The sample size was estimated according to Cochran's formula.

For this purpose, three steps were taken. The first step was a literature review for the study of barriers. The second step was developing a questionnaire. The third step was conducting a case study. The developed questionnaire was first distributed among the members of a pilot group for revision.

5.4.2 Selection of a List of Barriers Applicable to Iran

The first list of barriers was prepared with the study of the previous studies, which was very useful. It is noteworthy that the barriers listed in each study were specific to the region or country under study. Therefore, the barriers common in studies were selected. As was mentioned earlier, the questionnaire was first piloted to see whether the identified barriers apply to Iran.

In the first step, a panel of subject matter experts was used to measure the validity of the research tool. The results showed a high validity. Also, a pilot test was conducted exactly with the same questionnaire and categories, to test the reliability of the questionnaire. To check the reliability, Cronbach's alpha was calculated for each category and the result for each of them was above 0.7, which indicates the high reliability of the research tool. The participants were experts in the area of clean energy technologies excluded from the samples. The results showed high validity and reliability of the research tool. Finally, 28 barriers were identified which are listed in Table 5.1.

Table 5.1 The main barriers to the development of renewable

Category	Barriers	References
Institutional and regulatory barriers	Government's emphasis on the traditional fuel sector, lack of political stability, inadequate governmental support, lack of professional association, lack of NGO participation, and lack of adequate planning in the field of renewables.	Zhang et al. (2012) Nasirov et al. (2015) Luthra et al. (2015) Karatayev et al. (2016)
Economic and financial barriers	Initial investment costs, insufficient budget for project implementation, financial resource shortage, high subsidized energy supplied by the government, high transaction expenses, lack of financial mechanisms to encourage the use and development of clean energy, and low economic efficiency.	Reddy & Painuly (2004) Nasirov et al. (2015) Luthra et al. (2015) Karatayev et al. (2016)
Social and public awareness and information barriers	Poor awareness about the importance of energy, absence of consumer acceptance, absence of consumers' awareness of technology, lack of RE education and awareness programs for businesses and consumers, lack of social acceptance for some RETs, and absence of the contribution of stakeholders to policymaking.	Nasirov et al. (2015) Luthra et al. (2015) Karatayev et al. (2016) Carter et al. (2018)
Technological and infrastructure barriers	Shortage of local infrastructure, deficiency in R&D culture, lack of service provider, operation and maintenance of renewable energy equipment, lack of training facilities, lack of skilled personnel, technology complexity, lack of experts, nonexistence of standards, codes, and certification, and some inefficient technology.	Reddy & Painuly (2004) Nasirov et al. (2015) Luthra et al. (2015) Karatayev et al. (2016) Carter et al. (2018)

The statistical populations of the research were 68 experts of the RE Organization of Iran (SATBA), and 44 of them were selected randomly. The structure of the research questionnaire was based on respondents' opinions on the importance of the mentioned barriers. Initially, the respondents were given information about the objectives and structure of the research. Next, respondents were asked to rate the relative importance of each barrier on a 5-point Painuly scale. (Painuly, 2001). In addition, they also had the option to add their comments. The rates were from 1 to 5, namely, 1: Very less important to 5: Extremely important. Based on these rankings, the relative weight of each category was estimated using a formula as follows:

$$Wj = \frac{aj}{\Sigma aj}$$

Where *aj* is the average weight of the category *j* and Σaj is the relative weight of the category (Prasetiamartati et al., 2006).

5.5 Results and Discussion

The answers to the questionnaire provide useful information about the relative importance of each of the 28 barriers, which can be a good guide for government policymaking priorities for project implementation. The average score and ranking of the barriers are condensed in Tables 5.2 and 5.3. The mean score ranges from 4.18 to 2.97, with an overall average of 3.63, showing that all these obstacles are somehow applicable. The most important barriers in each category include "state focus on the fossil fuel sector", with the highest average value of 4.18, followed by "high initial investment costs" (4.15), "poor awareness about the clean energy" (3.81), and "lack of local infrastructure" (3.95). These important barriers in each category will be discussed further in the following sub-sections.

Table 5.2 Main barriers to the deployment of renewable energy technologies in Iran

Category	Barrier	Number of responses					Average score	Ranking	Standard deviation
		1	2	3	4	5			
Institutional and regulatory barriers	The government's emphasis on the traditional fuel sector	1	0	8	16	19	4.18	1	0.896
	Lack of political stability	1	3	8	22	10	3.84	2	0.938
	Insufficient legislative support and improper management	1	2	9	25	7	3.79	3	0.851
	lack of professional association	2	2	10	23	7	3.70	4	0.954
	Absence of NGO participation	1	3	11	24	5	3.65	5	0.861
	Lack of adequate planning in the field of renewables	1	4	10	23	6	3.65	5	0.913
Economic and financial barriers	Initial investment costs	0	1	5	24	14	4.15	1	0.713
	Insufficient funds to conduct projects	0	2	8	23	11	3.97	2	0.792
	Lack of financial resources	1	1	8	23	11	3.95	3	0.861
	High subsidized energy supplied by the government	1	2	7	24	10	3.90	4	0.884
	High transaction costs	1	0	13	25	5	3.75	5	0.750
	Lack of financial mechanisms to encourage and support the use and development of renewable energy	1	4	9	22	8	3.72	6	0.949
	Low economic efficiency	2	6	23	9	4	3.15	7	0.938

(Continued)

Table 5.2 *(Continued)*

Category	Barrier	Number of responses					Average score	Ranking	Standard deviation
		1	2	3	4	5			
Social and public awareness and information barriers	Poor knowledge about the importance of energy	0	5	8	21	10	3.81	1	0.921
	Lack of consumer acceptance of the product	2	4	6	22	10	3.77	2	1.053
	Absence of consumer awareness of technology	1	2	12	23	6	3.70	3	0.851
	Lack of RE education and awareness programs for businesses and consumers	1	5	6	28	4	3.65	4	0.887
	Lack of social acceptance for some RETs	1	6	22	11	4	3.25	5	0.892
	Absence of contribution of stakeholders in policymaking	4	4	20	14	2	3.13	6	0.978
Technological and infrastructure barriers	Lack of local infrastructure	0	3	7	23	11	3.95	1	0.834
	Deficiency in R&D culture	1	2	8	28	5	3.77	2	0.803
	Lack of service provider, operation, and maintenance of renewable energy equipment	1	2	11	25	5	3.70	3	0.823
	Lack of training facilities	1	6	20	10	7	3.36	4	0.990
	Lack of skilled personnel	0	5	23	12	4	3.34	5	0.805
	Technology complexity	0	4	26	11	3	3.29	6	0.733
	Lack of experts	2	7	17	14	4	3.25	7	0.991
	Lack of standards, codes, and certification	3	4	24	12	1	3.09	8	0.857
	Some inefficient technologies	2	8	24	9	1	2.97	9	0.820

Table 5.3 Ranking of barriers

Barriers' category	Rank and Responses 1	2	3	4	5	Average score	Weighted score	Rank	Sig
Institutional and regulatory barriers	7	14	56	133	54	3.80	0.261	1	0.036
Economic and financial barriers	6	15	68	126	49	3.79	0.260	1	
Social and public awareness and information barriers	9	26	74	119	36	3.55	0.243	2	
Technological and infrastructure barriers	10	41	160	144	41	3.41	0.234	3	

5.5.1 Institutional and Regulatory Barriers Category

In this category, according to the barriers scores, the Government's emphasis on the traditional fuel sector was the most important barrier; it should be mentioned that none of the previous studies around the world encountered such a critical obstacle. Experts believe that due to the abundance and availability of fossil fuels at the national level, the priorities of the Iranian government are focused on exploiting this resource for home consumption and exports. Iran is the largest government subsidizing fossil fuels, with $84 billion (or $1,083 per person) in 2013. These subsidies amount to more than one-fifth of Iran's gross domestic product, which is a significant figure for any country. Iran was the first country in the Middle East that made important improvements to its system of traditional fuel subsidization. The modification plan aimed to bring domestic traditional fuel prices close to international levels over five years. For example, the Targeted Subsidies Reform Act aimed to raise gasoline, diesel, kerosene, and LPG prices to 90% of Persian Gulf FOB prices. The first stage of the package involved noticeably increasing the domestic prices of petroleum products, natural gas, and electricity (Bárány & Grigonytė, 2015). Note that the

Institutional and Regulatory barrier's score was 3.8 and had the worst position among all other obstacles.

5.5.2 Economic and Financial Category

In this category, high initial investment costs had the highest average score among other barriers. Due to the high levels of primary investment required in these projects, access to primary funding is critical to developing projects using such technologies. Experts also mentioned that enormous investment expenditures, including primary and maintenance costs, high financial risks, and low returns in the early years, make people reluctant to install renewable energy technologies. It should be mentioned that the technology imported from extremely industrialized countries is more efficient, but at the same time, it may be costlier than technology manufactured locally, demanding higher initial investment costs. Primary costs for clean energy technologies tend to be extreme which may prohibit the consumers from accepting them. For many consumers, keeping the initial cost low is more important than minimizing operating costs; these findings are similar to the findings of previous studies (Hirmer & Cruickshank, 2014; Luthra et al., 2015; Reddy & Painuly, 2004). The funding options for RE plans in Iran, especially the role of the micro-credit sector, are very limited. Financial organizations in Iran are still very immature in the RE industry and they are reluctant to invest in large-scale RE projects. One of the most important barriers to financing these projects is the special characteristics of Iranian banks, including their conservative and risk-averse regulations. These findings are consistent with the findings of previous studies (Nasirov et al., 2015; Latin American Energy Organization, 2015). As shown in Table 5.3, the Economic and Financial barrier's score was 3.79, which shows the category's importance.

5.5.3 Social and Public Awareness and Information Category

Among the Social and Public Awareness and Information categories, poor knowledge about the importance of energy had the most important role in the barrier to the expansion of clean energy. This is due to Iran's lack of community capacity for information distribution. There is hardly any knowledge management system (software and/or hardware) about RETs (renewable energy technologies) that is freely accessible and simply reachable for the stakeholders. Under these conditions, gathering, and processing information is time-consuming for consumers, and this is one of the main difficulties of using renewable technologies encountered by the public. These results are in line with the results of a study by Mondal et al. (2010). Information diffusion associated with clean energy should embrace everything from government incentives to support systems. Lack of knowledge of technology performance and management and limited access to spare parts and the necessary expertise to repair and maintain devices are the few difficulties in accepting clean energy technologies; these findings are similar to the findings of previous studies (Sawin, 2003; Karakosta et al., 2010; Carter et al., 2018). In addition, the Social and Public Awareness and Information barrier average score was 3.55, which shows the category's importance.

5.5.4 Technological and Infrastructure Category

In this category, the lack of local infrastructure had the most significant role. Experts believe that Iran's infrastructural capacity is not enough. Therefore, changes in operational, managerial, and flexibility patterns in infrastructures should be made to increase the share of renewables. Iran ranked 84th among 110 countries in terms of loss of electricity in the distribution network in 2010. Statistics show that the total electricity loss in transmission and distribution networks in 2011 in Iran was 18.1%, and in 2012, it increased to 18.7%. Losing 1% means losing about $300 million of fuel at power plants. These results are in line with the findings of Afsharzade et al. (2016). It should be mentioned that according to the findings, the Technological and Infrastructure barrier's score was calculated to be 3.41, which shows the category's importance. The results are presented in Tables 5.2 and 5.3. We can see that among the different categories of mentioned barriers, the Institutional and Regulatory category was the highest among others (weighted score = 0.261), followed by the

Economic and Financial (weighted score = 0.260), Social and Public Awareness and Information (weighted score = 0.243), and Technological and Infrastructure (weighted score = 0.234), respectively. Graphically, different categories of obstacles to the positioning of clean energy sources in Iran can be seen in Fig. 5.10. The results of the weighted score mean that based on experts' judgment, Institutional, and Regulatory Barriers had the worst condition in Iran; however, no previous studies have pointed to the strong role of these barriers, and this is due to the specific policymaking conditions in Iran.

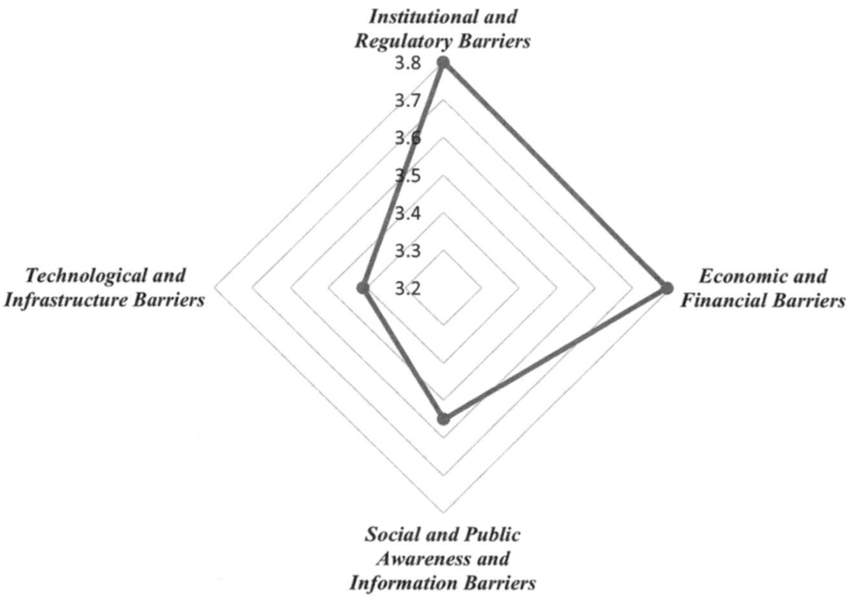

Figure 5.10 Kite diagram of barriers to the development of renewable energies.

According to Fig. 5.11, all categories are above 3, and it shows that all of them are important.

For comparing differences in average scores and weighted scores as important in determining the importance and relative rank of the different barrier categories, the Kruskal–Wallis test was used. The results indicate that there is a significant difference between different categories of challenges at a significant level of 95% (Fig. 5.11).

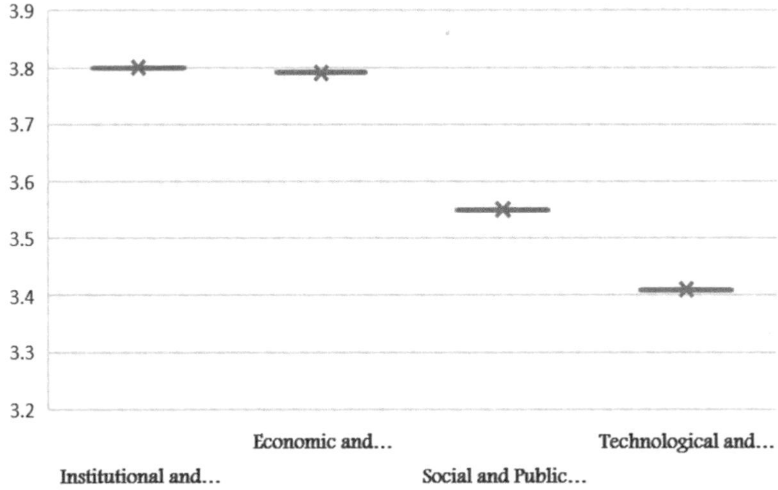

Figure 5.11 Status of barriers.

5.6 Conclusion and Policy Implications

Key barriers include a government focus on the traditional fuel sector, high initial investment costs, poor awareness about the importance of energy, and lack of local infrastructure. Until 2050, RE penetration requires a set of actions at the national, regional, and local levels to be considered. The scenarios for considerable growth in clean energy establishment in Iran are challenging, particularly when oil price fluctuations and sanctions continue to damage Iran's economic situation. If parts of the oil export revenues or fossil fuel subsidies are allocated to expand RE programs, Iran will be saved from dependence on hydrocarbon fuels and ecological pressures. It is necessary to consider institutional, economic, social, and technological dimensions in the selection of these policies to achieve a sustainable future.

In addition, the government's measures in offering credit warrants or supplying "green bonds" for domestic and foreign profitable banks may aid in assembling long-term funding for the advancement of renewable sources. Furthermore, as a segment of a government strategy, it is vital to inspire energy-intensive trades in the country to assign funds for the advancement of industrial innovation and renewable pilot plans. Similarly, as another choice, in the primary stage of expansion, international economic organizations may have a significant function to show in funding RET plans and particularly in speeding up market linkages. The financing flow from organizational investors may send a good signal of investors' assurance and knowledge to the home-grown banks, which should simplify the payment of loans for further plans.

It appears that pioneering policymaking is necessary to deal with the barriers to the non-proliferation of RE development. Regarding challenges for RE development, a mixture of some policy mechanisms is essential for the penetration of RE. These policy mechanisms include phasing out traditional fuel aids to fund the needed incentive mechanisms for clean energy; setting up more inducement mechanisms for RET's development; raising public awareness about undesirable consequences of fossil fuel utilization by education and information broadcasting systems to expand social acceptance; growing straight financing in R&D; identification of suitable areas in the country to exploit RE sources by providing information and maps of wind energy (areas prone to permanent and high-intensity winds), solar energy (areas with high radiation intensity and high exposure time of sunlight during the day), geothermal energy (geographical areas where geothermal energy is recoverable with minimal investment), and biomass energy (areas where production and consumption of energy from biomass is available and cheap); and feasibility study for RET's development.

For future studies, considering the identification of the "Institutional and Regulatory Barriers" as the most important barrier and a study on "Designing an optimal policymaking model for renewable energies in Iran" are suggested.

Also, some inefficient technologies were one of the experts' concerns. To inform cost-benefit analysis is used to evaluate energy efficiency measures as a general approach. These evaluations can be derived from different analytical perspectives, e.g., from a social or governmental position otherwise from a planning or an investor point of view.

Therefore, it is suggested to analyze this issue in detail from different perspectives in future studies.

Although this study has provided insight into the challenges preventing the development of RE technology in Iran, there are still certain gaps in the literature that make it necessary for further study. Further research should encompass various critical aspects of advancing clean energy technologies. This includes examining the latest technological advancements and their adaptability to local contexts, conducting comprehensive economic viability studies for clean energy projects while examining diverse financing models, and ensuring the development of sustainable financial structures. In addition, an in-depth analysis of existing policy and regulatory frameworks should identify gaps and inconsistencies to aid the transition to RE sources. Understanding how social and cultural factors influence public perception and acceptance of clean energy technologies at the community level is crucial. Moreover, assessing the current energy infrastructure, addressing challenges related to grid integration and stability, and exploring environmental impact assessments are essential steps. Lastly, recognizing regional variations is vital to exploring unique obstacles and opportunities for clean energy adoption in different geographical areas and under varying local economic dynamics. These research activities contribute to a greater awareness of clean energy issues and possibilities, which aids in the development of specialized policies to promote sustainable energy and effectively address environmental and energy security challenges.

In conclusion, it is essential to recognize and describe its inherent limitations. Firstly, the research faced constraints related to the level of public familiarity with various RE technologies. This unfamiliarity among the general population may have resulted in a somewhat limited understanding of the challenges associated with clean energy development from their perspective. As a consequence, there could be unexplored dimensions of these challenges that have not been fully explained in this study. Moreover, the study's scope and depth may have been influenced by factors such as time constraints, resource availability, and the accessibility of certain data sources. These practical limitations may have prevented a more exhaustive examination of the subject matter. Additionally, the study's findings are based on the data collected during a specific timeframe, and the dynamic nature of energy-related developments implies that new challenges and opportunities may have emerged since then. Furthermore, the research may have encountered limitations associated with sample size or selection bias, which could impact the generalizability of the findings to the broader population. While this study has provided valuable insights, it is essential to recognize these limitations because they underscore the need for future research endeavors to delve deeper into the multifaceted challenges of clean energy adoption. Additionally, it is important to consider more extensive engagement strategies to bridge the knowledge gap among the public.

References

Abas, A. R., Tadayon, S. M., & Yasin, H. M. (2021). Renewable energy integration challenges in Iran: A techno-economic review. *Energy Strategy Reviews*.

Abdoli, M. (1966). Biogas, Atomic Energy Organization of Iran.

Adrianto, L., Matsuda, Y., & Sakuma, Y. (2005). Assessing local sustainability of fisheries system: A multi-criteria participatory approach with the case of Yoron Island, Kagoshima prefecture. *Japan Marine Policy*, 29(1), 9–23.

Afful-Dadzie, A., Afful-Dadzie, E., Abbey, N.A., Owusu, B.A., & Awudu, I. (2020). Renewable electricity generation target setting in developing countries: Modeling, policy, and analysis. *Energy for Sustain Dev*, 59, 83–96.

Afsharzade, N., Papzan, A., Ashjaee, M., Delangizan, S., Van Passel, S., & Azadi, H. (2016). Renewable energy development in rural areas of Iran. *Ren Sustain Energy Rev*, 65, 743–755.

Agarwal, A.K. (2007). Biofuels (alcohols and biodiesel) applications as fuels for internal combustion engines. *Prog Energy Comb Sci*, 33(3), 233–271.

Akbi, A., Saber, M., Aziza, M., & Yassaa, N. (2017). An overview of sustainable bioenergy potential in Algeria. *Ren Sust Energy Rev*, 72, 240–245.

Aslani, A., Naaranoja, M., & Antila, E. (2012). The feasibility study of prior renewable energy alternatives to private sector investment. *Int J Adv Renew Energy Res*, 1(5), 248–53.

Azadi, P., Sarmadi, A. N., Mahmoudzadeh, A., & Shirvani, T. (2017). The outlook for natural gas, electricity, and renewable energy in Iran. *Stanford Iran*, 2040, 1–27.

Bárány, A., & Grigonytė, D. (2015). Measuring fossil fuel subsidies ECFIN. *Econ Brief*, 40, 1–13.

BP Statistical Review of World Energy (2014). Excel workbook of historical data, 2014.

Carter, E., Shan, M., Zhong, Y., Ding, W., Zhang, Y., Baumgartner, J., & Yang, X. (2018). Development of renewable, densified biomass for household energy in China. *Energy for Sust Dev*, 46, 42–52.

Cheraghi, S., Choobchian, S., & Abbasi, E. (2019). Investigation of entrepreneurship development barriers in the field of renewable energies technologies in developing countries: A case of Iran. *Int J Sci Technol Res*, 8(3), 11.

Cheraghi, S., Choobchian, S., & Abbasi, E. (2020). Impact of institutional pressure on the investment decision-making process in renewable energy technologies in the agricultural sector, Iran. *Plant Archives*, 20(1), 1743–1751.

Dehhaghi, S., & Choobchian, S. (2020). Renewable energy, the key to sustainability in food production. Paper presented at the 7th Conference of Applied Researches in Agricultural Sciences.

Dehhaghi, S., Choobchian, S., Ghobadian, B., & Farhadian, H. (2022). Policy model of renewable energy development in Iran's agriculture sector. *Renewable Energy and Environment*, 9(2), 93–106.

Dehhaghi, S., Choobchian, S., Ghobadian, B., Farhadian, H., Viira, A.-H., Stefanie, H. I., & Azadi, H. (2022). Five-year development plans of renewable energy policies in Iran: A content analysis. *Sustainability*, 14(3), 1501.

Dyer, J.A., & Desjardins, R.L. (2018). Energy use and fossil CO_2 emissions for the Canadian fruit and vegetable industries. *Energy for Sustain Dev*, 47, 23–33.

Eslamian, S. S., Khedmati, M. K., & Marzbani, M. S. (2018). Geothermal energy prospects in Iran: A geographical assessment. *Renewable and Sustainable Energy Reviews*.

Energy Balances (2011). Power and Energy Planning Department, Ministry of Energy of I.R. Iran.

Energy Balance Sheet (2014). Deputy of electricity and energy affairs, Tavanir Expert Holding Company, Tehran.

European Environment Agency (2015). Greenhouse gas emissions per capita and unit of GDP in purchasing power standards. Available online: http:// www.eea.europa.eu/data.

Fadai, D., Esfandabadi, Z.S., & Abbasi, A. (2011). Analyzing the causes of non-development of renewable energy-related industries in Iran. *Ren Sus Energy Rev.*, 15(6), 2690–2695.

Ghouchani, M., Taji, M., Cheheltani, A. S., & Chehr, M. S. (2021). Developing a perspective on the use of renewable energy in Iran. *Technological Forecasting and Social Change*, 172, 121049.

Ghorashi, A.H., & Rahimi, A. (2011). Renewable and non-renewable energy status in Iran: Art of know-how and technology-gaps. *Ren Suste Energy Rev*, 15(1), 729–736.

Hassanzadeh, E. (2012). Recent developments in Iran's energy subsidy reforms, Policy Brief. *Int Ins for Sust Dev*, 1–12. Available at: http:// www.iisd.org/gsi/sites/default/files/pb14_iran.pdf.

Hirmer, S., & Cruickshank, H. (2014). Making the deployment of pico-PV more sustainable along the value chain. *Ren Sust Energy Rev*, 30, 401–411.

Hosseini, S.E., Andwari, A.M., Wahid, M.A., & Bagheri, G. (2013). A review on green energy potentials in Iran. *Ren Sust Energy Rev*, 27, 533–545.

Intergovernmental Panel on Climate Change (2014). *Mitigation of Climate Change*, Cambridge University Press. https://www.ipcc.ch/.

International Energy Agency (2011). World Energy Model Methodology and Assumptions.

International Energy Agency (2015). Energy technology perspectives: Mobilizing innovation to accelerate climate action.

International Energy Agency (1997). Key issues in developing renewable in Paris.

International Energy Agency (2014). Available from: https://www.iea.org/.

Iran Renewable Energy Organization (2010). http://www.suna.org.ir/home-en. html.

Jahangiri, M., Nematollahi, O., Haghani, A., Raiesi, H. A., & Alidadi Shamsabadi, A. (2019). Optimization of energy cost of clean hybrid solar-wind power plants in Iran. *International Journal of Green Energy*, 16(15), 1422–1435.

Karakosta, C., Doukas, H., & Psarras, J. (2010). Technology transfer through climate change: Setting a sustainable energy pattern. *Ren Sust Energy Rev*, 14(6), 1546–1557.

Karatayev, M., Hall, S., Kalyuzhnova, Y., & Clarke, M.L. (2016). Renewable energy technology uptake in Kazakhstan: Policy drivers and barriers in a transitional economy. *Ren Sus Energy Rev*, 66, 120–136.

Khoiyangbam, R.S. (2011). Environmental implications of biomethanation in conventional biogas plants. *Iran J Energy Environ*, 2(2), 181–187.

Laha, P., & Chakraborty, B. (2017). Energy model–a tool for preventing energy dysfunction. *Ren Sust Energy Rev*, 73, 95–114.

Latin American Energy Organization (2015). United Nations Industrial Development Organization, Observatory of Renewable Energy in Latin America and Caribbean, Chile, Final Report, Component 3: Financial Mechanism. Available online: http://www.renenergyobservatory.org/uploads/media/ChileProducto3Eng07.pdf.

Li, M., & Wang, Q. (2017). Will technological advances alleviate climate change? Dual effects of technology change on aggregate carbon dioxide emissions. *Energy for Sustain Dev*, 41, 61–68.

Luthra, S., Kumar, S., Garg, D., & Haleem, A. (2015). Barriers to renewable/sustainable energy technologies adoption: Indian perspective. *Ren Sust Energy Rev*, 41, 762–776.

Mahdavi, A., & Yarahmadi, R. (2014). Iran's wind energy landscape: Current status and future prospects. *Renewable and Sustainable Energy Reviews.*

Mohammadi Maghanaki, M., Ghobadian, B., Najafi, G., & Galogah, R.J. (2013). The potential of biogas production in Iran. *Ren Sust Energy Rev*, 28, 702–714.

Momenpour, Y., & Choobchian, S. (2022). A look at the environmentalist behavior of farmers in the water-food-energy link. *The 9th National Science Congress of Agricultural Extension and Education, Natural Resources and Sustainable Environment*, September 17 (Tabriz University).

Farangi, M., Soleimani, E. A., Zahedifar, M., Amiri, O., & Poursafar, J. (2020). The environmental and economic analysis of grid-connected photovoltaic power systems with silicon solar panels, in accord with the new energy policy in Iran. *Energy*, 202, 117771.

Mendoza, G.A., & Prabhu, R. (2003). Qualitative multi-criteria approach to assessing indicators of sustainable forest resource management. *Forest Ecol Manage*, 174(1–3), 329–343.

Ministry of Agriculture Jihad (2019). Annual Report 2019, Ministry of Agriculture Jihad: Tehran, Iran.

Mohamadi, H., Saeedi, A., Firoozi, Z., Zangabadi, S. S., & Veisi, S. (2021). Assessment of wind energy potential and economic evaluation of four wind turbine models for the east of Iran. *Heliyon*, 7(6).

Mohammadi Maghanaki, M., Ghobadian, B., Najafi, G., & Galogah, R.J. (2013). The potential of biogas production in Iran. *Ren Sust Energy Rev*, 28, 702–714.

Mondal, M.A.H., Kamp, L.M., & Pachova, N.I. (2010). Drivers, barriers, and strategies for implementation of renewable energy technologies in rural areas in Bangladesh—An innovation system analysis. *Energy Policy*, 38(8), 4626–4634.

Mostafaeipour, A., Alvandimanesh, M., Najafi, F., & Issakhov, A. (2021). Identifying challenges and barriers for development of solar energy by using fuzzy best-worst method: A case study. *Energy*, 226, 120355.

Najafi, G., Ghobadian, B., Mamat, R., Yusaf, T., & Azmi, W.H. (2015). Solar energy in Iran: Current state and outlook. *Ren Sust Energy Rev*, 49, 931–942.

Najafi, G., & Ghobadian, B. (2011). LLK1694-wind energy resources and development in Iran. *Ren Sust Energy Rev*, 15(6), 2719–2728.

Najafi, G., & Ghobadian, B. (2011). Geothermal resources in Iran: The sustainable future. *Ren Sust Energy Rev*, 15(8), 3946–3951.

Nasirov, S., Silva, C., & Agostini, C.A. (2015). Investors' perspectives on barriers to the deployment of renewable energy sources in Chile. *Energies*, 8(5), 3794–3814.

Noorani, A., Rahimi, M., & Mirbagherijam, M. (2022). Survey the effect of types of renewable and non-renewable energies consumption on economic welfare in Iran. *Journal of Renewable and New Energy*, 9(1), 61–68.

Oil & Gas (2015). Worldwide look at reserves and production.

Oryani, B., Koo, Y., Rezania, S., & Shafiee, A. (2021). Barriers to renewable energy technologies penetration: Perspective in Iran. *Renewable Energy*, 174, 971–983.

Painuly, J.P. (2001). Barriers to renewable energy penetration: A framework for analysis. *Ren Energy*, 24(1), 73–89.

Prasetiamartati, B., Fauzi, A., Dahuri, R., Fahrudin, A., & Lange, H. (2006). Destructive fishery and fishery sustainability assessing fishery sustainability using a multicriteria participatory approach: A case study of small islands in South Sulawesi. *J Cost Dev Opm*, 9(3), 163–174.

Reddy, S., & Painuly, J.P. (2004). Diffusion of renewable energy technologies—Barriers and stakeholders' perspectives. *Ren Energy*, 29(9), 1431–1447.

Rezaei, M., Chaharsooghi, S.K., & Abbaszadeh, P. (2013). The role of renewable energies in sustainable development: Case study Iran. *Iran J Energy Environ*, 4(4), 320–329.

Richards, G., Noble, B., & Belcher, K. (2012). Barriers to renewable energy development: A case study of large-scale wind energy in Saskatchewan, Canada. *Energy Policy*, 42, 691–698.

Rosen, M.A. (2009). Energy, environmental, health and cost benefits of cogeneration from fossil fuels and nuclear energy using the electrical utility facilities of a province. *Energy for Sustain Dev*, 13(1), 43–51.

Sansaniwal, S. K., Pal, K., Rosen, M.A., & Tyagi, S.K. (2017). Recent advances in the development of biomass gasification technology: A comprehensive review. *Renewable and Sustainable Energy Reviews*, 72, 363–384.

SATBA. Report 4: Biomass Energy 1; SATBA: Tehran, Iran, 2005. Available online:http://www.dres-isf.ir/home/phocadownload/%D8%B2%DB%8C%D8%B3%D8%AA%20%D8%AA%D9%88%D8%AF%D9%87.pdf (accessed on 25 June 2021).

Sawin, J. (2003). *Charting a New Energy Future*. State of the world in: Starke Linda, editor. New York: W.W. Norton & Company, 85–109.

Shokri, S. (2012). Biogas technology, applications, perspectives and implications. *Int J Agri Sci Res*, 2(3), 53–60.

Sindhu, S., Nehra, V., & Luthra, S. (2016). Identification and analysis of barriers to implementation of solar energy in the Indian rural sector using integrated ISM and fuzzy MICMAC approach. *Ren Sus Energy Rev*, 62, 70–88.

Trends in Global CO_2 emissions. (2015). Available online: http://edgar.jrc.ec.europa.eu/news_docs/jrc-2015-trends-in-global-co$_2$-emissions-2015-report-98184.pdf.

United Nations Development Programme (2007). Fighting climate change: Human solidarity in a divided world. Palgrave Macmillan UK. Retrieved from: Http://hrd.undp.org.

United Nations Environment Programme (2015). Policy Coherence of the Sustainable Development Goals: A Natural Resource Perspective.

U.S. Energy Information Administration (2015). Iran Country Analysis Brief. Available from: http://www.eia.gov/beta/international/analysis.cfm?iso=IRN.

U.S. Energy Information Administration (2018). https://www.eia.gov/beta/international/an alysis.php ?iso=IRN.

World Energy Outlook (2017). http://www.worldenergyoutlook.org/weomodel.

World Energy Outlook (2015). International Energy Agency. Available: www.worldenergyoutlook.org.

Yaghoubi, J., Yazdanpanah, M., & Komendantova, N. (2019). Iranian agriculture advisors' perception and intention toward biofuel: A green way toward energy security, rural development, and climate change mitigation. *Renewable Energy*, 130, 452–459.

Yazdanpanah, M., Komendantova, N., & Zobeidi, T. (2022). Explaining intention to apply renewable energy in agriculture: The case of broiler farms in Southwest Iran. *International Journal of Green Energy*, 19(8), 836–846.

Zahedi, R., Zahedi, A., & Ahmadi, A. (2022). Strategic study for renewable energy policy, optimizations and sustainability in Iran. *Sustainability*, 14(4), 2418.

Zare, N. A., Haghifam, M. R., & Kazemi, A. (2020). Clean energy transition in Iran: A case study of solar photovoltaic deployment. *Energy Policy*.

Zhang, X., Shen, L., & Chan, S.Y. (2012). The diffusion of solar energy uses in HK: What are the barriers? *Energy Policy*, 41.

Chapter 6

Solar Enhanced Soaring: Tapping into Ambient Solar Energy for Optimized Airfoil Performance*

Mohd Sahil Akhtar,[a] David S.-K. Ting,[b] and Jagannathan T. Kalathi[c]

[a]*Soft Materials Research Laboratory, Department of Chemical Engineering, National Institute of Technology Karnataka, Surathkal, Mangalore 575025, Karnataka, India*
[b]*Turbulence and Energy Laboratory, University of Windsor, Windsor, Ontario, Canada*
[c]*Soft Materials Research Laboratory, Department of Chemical Engineering, National Institute of Technology Karnataka, Surathkal, Mangalore 575025, Karnataka, India*

jagantk@nitk.edu.in

The amount of solar and background radiation absorbed by birds varies according to their wing shape, pigmentation, porosity, etc. Birds are equipped with unique features to thrive, including attracting the opposite sex, regulating body temperatures, and soaring in the sky. The research focuses on solar/sky radiation

*This chapter is based on the article "Bio-inspired study of thermal effects on NACA0012 airfoil at Reynold's Number of 33,000", published at Leddy Library, University of Windsor, where the authors retain 100% copyright.

Reclaiming Eden: Responsible Living, Engineering, and Architectures
Edited by David S.-K. Ting and Jacqueline A. Stagner

ISBN 978-981-5129-31-1 (Hardcover), 978-1-003-49612-0 (eBook)
www.jennystanford.com

by examining how the NACA0012 airfoil, representing the wing of a bird, performs when its upper surface temperature is higher or lower than the surrounding ambient air. This is realized by performing 2-dimensional simulations in OpenFOAM at a Reynolds number of 33,000, where the Spalart-Allmaras model is used to simulate the flow turbulence. The upper surface of the airfoil is warmed to 330 K and cooled to 270 K at a pressure of 1 atm, an ambient temperature of 300 K, and a Mach number of 0.0725. The results illustrate the airfoil with the cooler top surface exhibits a lower drag and higher lift than its warmer top surface counterpart. A maximum reduction of drag coefficient from 0.065 to 0.061 and an increase in lift coefficient from 0.89 to 0.93 at an angle of attack of 11° are achieved. In short, tuning the upper surface of the NACA0012 airfoil to temperatures lower than the ambient provides better aerodynamic performance.

Nomenclature

A	Area of airfoil
AOA	Angle of attack
c	Chord length of the airfoil
C_D	Drag coefficient
C_L	Lift coefficient
CM	Airfoil control mass
CV	Airfoil control volume
D	Drag force acting on the airfoil
e	Total energy per unit mass
f	Net force acting on a body
f_b	Body force
f_s	Surface force
L	Lift force acting on the airfoil
m	Mass of airfoil control mass
Ma	Mach number
p	Pressure
$\dot{q}_s$	Rate of heat transfer per unit area
$\dot{q}_v$	Rate of heat source/sink per unit volume

R	Gas constant
RANS	Reynolds Averaged Navier-Stokes
Re	Reynolds number
S	Airfoil control surface
T, ΔT	Temperature, temperature difference
T_a, T_l, T_u	Ambient, lower, and upper temperature
v	Free stream velocity
δ	Thickness of airfoil
V	Volume of control volume
μ	Viscosity
ρ	Density

6.1 Introduction

There has been a lot of focus on drone design in the past several years. Their essential qualities, including effortless deployment, adjustable altitude, flexibility, and mobility, make them suitable for a wide range of practical applications [1, 2]. Their versatility signals the beginning of a new era in which unmanned aerial vehicles, in addition to surveillance and remote sensing, are gaining ground in many important industries for non-destructive testing and the early detection of catastrophic failures [3]. Aside from improvements to existing designs, research has also advanced the development of unique drones of various shapes, sizes, and weights [4]. As a result, the configuration of drones varies greatly depending on the platform and requirements of the operation, and they are classified in various ways based on several factors [5]. The smaller aerial vehicles do, however, have one major weakness, like their power system is insufficient to support extended flying durations [6]. Finding an excellent alternative to increasing current unmanned aerial vehicles' aerodynamic performance is thus necessary.

Several natural occurrences help to alter the way that innovation is perceived. It is critical to harness these occurrences considering the present energy crisis [7]. The want for energy has prompted more studies into improving performance to optimizing effectiveness by using nature as a doorway. Such stimulus has led to fantastic advancements in the science of biomimicry and bio-inspiration, creating techniques for lowering drag in all forms of air and water transportation [8, 9]. Birds are highly effective flyers, having promising advantages in bio-inspired drone design [10]. Albatross and other large seabirds are known to tap into dynamic soaring to travel large distances with a fraction of the effort [11]. The question of why their flight is so efficient does not come as a surprise. It is a sum of varying unique characteristics collectively optimized that helps them surpass the performance of any manufactured vehicle of an equivalent size [12]. The stark difference between the ambient and wing temperature of seabirds has garnered attention to the coloration of their feathers, as shown in Fig. 6.1.

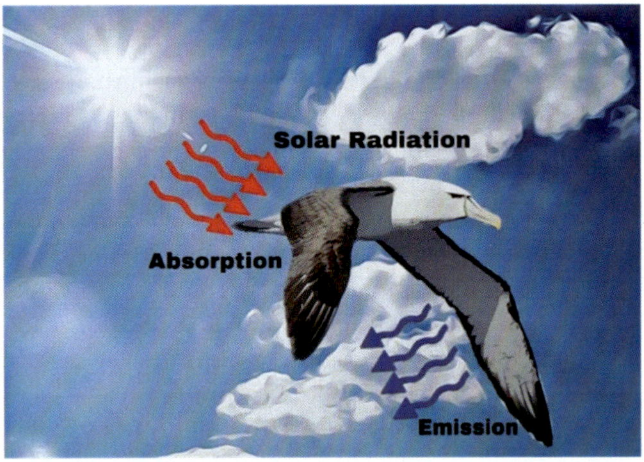

Figure 6.1 Black and white wing coloration of an albatross under solar radiation.

The wing pigmentation of birds such as albatrosses, shearwaters, sooty terns, and black swimmers impacts and helps reduce skin drag [13, 14]. Due to differences in solar absorptivity, the upper and lower surfaces of the wing, which are black and white, impact the amount of solar/sky energy absorbed from the ambient environment. Therefore, a temperature difference exists between the top and bottom wing surfaces relative to the ambient. Measurements of the maximum average temperature show that the black wing can be heated to as much as 81.3 °C while the white wing only heats to 53.1 °C [15]. Similarly, darker feathers warmed up to 70.7 °C in the absence of wind, which was 31.0 °C hotter than the brighter ones when exposed to solar radiation [16]. An experimental study on different airfoil types using a temperature difference of 10 °C between the wind tunnel ambient temperature and the top wing surface revealed a reduction in drag and lift improvement for almost all studied angle of attack (AOA) values [15]. The aerodynamic performance, measured in dimensionless lift and drag coefficients (C_L and C_D), are described in terms of airfoil reference area (A) and chord length (c), under the influence of a lift (L) and drag (D) force, in a freestream of velocity (v), density (ρ), and viscosity (μ):

$$C_L = \frac{L}{1/2(A\rho v^2)} \tag{6.1}$$

$$C_D = \frac{D}{1/2(A\rho v^2)}. \tag{6.2}$$

The values of C_D and C_L for different types of airfoils reported are provided in Table 6.1 [15] at $Re \cong 10^5$, where Re is the flow Reynolds number given by:

$$Re = \frac{\rho vc}{\mu}. \tag{6.3}$$

The results suggest that aerodynamic performance was improved by tapping into the ambient solar energy that helps maintain warmer upper wing surface temperature. The C_D values decreased for all airfoils, with a maximum percentage drop in C_D (ΔC_D = 50%) noticed for the GOE174 airfoil. The maximum improvement in C_L (ΔC_L = 71%) was observed for the E186 airfoil. The best performance was exhibited by the same airfoil, for which the C_L/C_D value increased by threefold [15].

Table 6.1 Summary of lift improvement and drag reduction by upper surface heating of 10 °C on different airfoils [15]

Airfoil	C_D w/o heating	C_D with heating	ΔC_D (%)	C_L w/o heating	C_L with heating	ΔC_L (%)	$\frac{C_L}{C_D}$ w/o heating	$\frac{C_L}{C_D}$ with heating	$\Delta\left(\frac{C_L}{C_D}\right)$%
NACA0012	0.14	0.11	21 ↓	0.19	0.22	16 ↑	1.61	2.05	27 ↑
GOE174	0.06	0.03	50 ↓	0.18	0.25	39 ↑	2.50	7.10	184 ↑
S5020	0.07	0.06	14 ↓	0.12	0.14	17 ↑	1.32	1.80	36 ↑
E186	0.13	0.07	46 ↓	0.21	0.36	71 ↑	0.70	2.12	203 ↑
MH20	0.09	0.07	22 ↓	0.08	0.12	50 ↑	0.91	1.61	77 ↑

Table 6.2 Summary of effects of temperature* warmer and cooler than ambient on the performance of different airfoils

Parameters	[17]	[18]	[19]	[20]	[21]
Airfoil	NACA0012-64	Pitching NACA0012	NACA0012	NACA0012	NACA2412
	750,000	3,000	33,000	38,000	3,000
Temperature	Wing-to-freestream temperature ratio increased from 1 to 2	$T_u = T_l = 350$ K $T_a = 300$ K	$T_u = 200$ K $T_l = 400$ K $T_a = 300$ K	$T_u = 200$ K $T_l = 400$ K $T_a = 300$ K	$T_u = 200$ K $T_l = 400$ K $T_a = 300$ K
C_L	Decrease from 1.00 to 0.70 ↓	Decrease from 0.86 to 0.75 ↓	Increase from 0.66 to 0.96 ↑	Increase from 0.85 to 1.17 ↑	Increase from 0.04 to 0.12 ↑
C_D	Increase from 0.006 to 0.009 ↑	Decrease from 0.37 to 0.35 ↓	Decrease from 0.120 to 0.062 ↓	Decrease from 0.130 to 0.082 ↓	Decrease from 0.073 to 0.071 ↓
Conclusion	Overall performance decreased	Overall performance decreased	C_L/C_D increase from 5.40 to 15.5	C_L/C_D increase from 6.64 to 14.27	C_L/C_D increase from 0.58 to 1.72

*The subscripts in T, namely, a, u, and l stand for ambient, upper, and lower surface temperature, respectively.

Table 6.2 highlights studies contrary to the results in Ref. [15]. At $Re = 7.5 \times 10^5$, it was revealed that C_L decreased from

1.0 to 0.7 for NACA0012-64 airfoil wall temperature warmer than ambient [17]. A study of NACA0012 at $Re = 3000$, $\Delta T = 50$ K, $T_a = 300$ K concluded a decrease in C_L from 0.86 to 0.76, and C_D decrease from 0.37 to 0.35 [18]. NACA0012 airfoil at $Re = 3.3 \times 10^4$ with $T_u = 200$ K and $T_l = 400$ K at $T_a = 300$ K concluded increase in C_L/C_D from 5.4 to 15.5. At $Re = 3.8 \times 10^4$ and thermal conditions same as in Ref. [19] concluded increase in C_L/C_D from 6.64 to 14.27 [20]. Ref. [21] studied NACA2412 at $Re = 3{,}000$ with $T_u = 200$ K, $T_l = 400$ K, and $T_a = 300$ K, and documented C_L/C_D increase from 0.58 to 1.72. In addition, it was found that at $Re = 7.5 \times 10^5$, the C_L value decreases from 1.30 to 1.15 with NACA0012-64 airfoil upper surface temperature twice the ambient [22]. The same methodology was implemented for applications in wind turbines. At $Re = 5.5 \times 10^5$, $T_u = 225$ °C and $T_l = 25$ °C, the C_L/C_D ratio decreased from 33.2 to 20.4 for S809 airfoil; whereas, with $T_u = 25$ °C and $T_l = 225$ °C, almost no change was observed in the airfoil performance [23].

The preceding section shows an existing disparity between the consequences of upper-surface warming and cooling. While some studies show that cooling the upper surface of an airfoil improves performance [18–21], Ref. [15] indicates otherwise.

Thus, in the present study, we aim to investigate the effects of warming and cooling the upper surface of the NACA0012 airfoil relative to ambient temperature. Figure 6.2 shows the difference between the NACA0012 airfoil and those used in bio-inspired studies that include NACA2412, MH20, S5020, GOE174, and E186 [15]. Thus far, GOE174 is believed to be the closest representation of the albatross wing, although E186 shows maximum improvement in performance with a warmer upper surface. There are inherent differences in the shape and structure of the geometry between the airfoil used in the present study and those used in bio-inspired research. However, investigating thermal effects on the upper surface of a basic airfoil forms the first step in developing a good understanding of the force and flow characteristics. Therefore, we performed 2-dimensional simulations on the NACA0012 airfoil in OpenFOAM using the Spalart-Allmaras turbulence model to capture the dynamics of turbulence. Since birds cannot sustain temperatures as high as

400 K and as low as 200 K, we chose the warmer and cooler temperatures of the upper surface within the bounds of the practical temperature difference between the upper black and lower white wings reported in Ref. [15, 16]. The resolution of the ambiguity mentioned above will help us better understand the fundamentals of aerodynamics and help us tap into further depths of energy optimization through the gateway of solar energy.

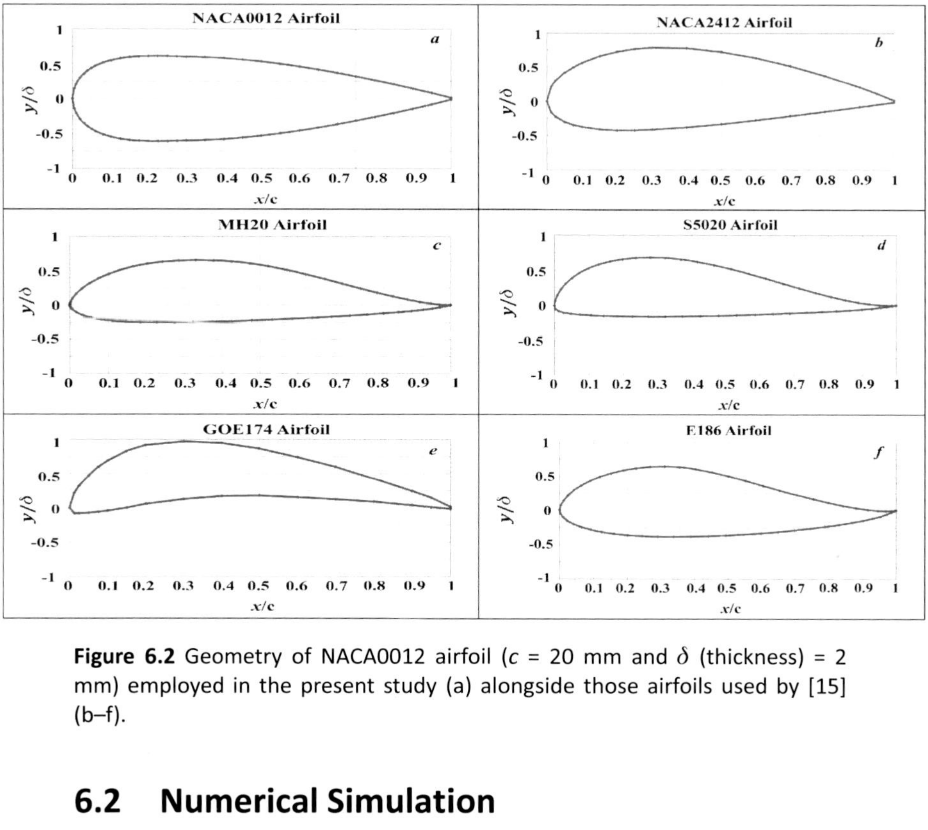

Figure 6.2 Geometry of NACA0012 airfoil (c = 20 mm and δ (thickness) = 2 mm) employed in the present study (a) alongside those airfoils used by [15] (b–f).

6.2 Numerical Simulation

The numerical simulation is based on the RANS Spalart-Allmaras method. The computation comprises (1) geometry of the 2D NACA0012 airfoil creation, (2) conceptualizing the flow governing equations, and (3) grid generation and validation of the simulation model.

6.2.1 Geometry of the Airfoil

Figure 6.3 shows the essential terminologies associated with an airfoil. The leading edge is the point at the front of the airfoil with maximum curvature. Similarly, the trailing edge is the point at the back of the airfoil with maximum curvature. The chord line joins the leading and trailing edge, denoted by the letter *c*. The mean camber line is the locus of points midway between the upper and lower surface of the airfoil. The maximum camber is the maximum difference between the mean camber line and the chord line. Finally, AOA is the angle made by the wind direction and the chord line of the airfoil. In the body of the paper, these terms will be mentioned again, with AOA being referred to in degrees throughout.

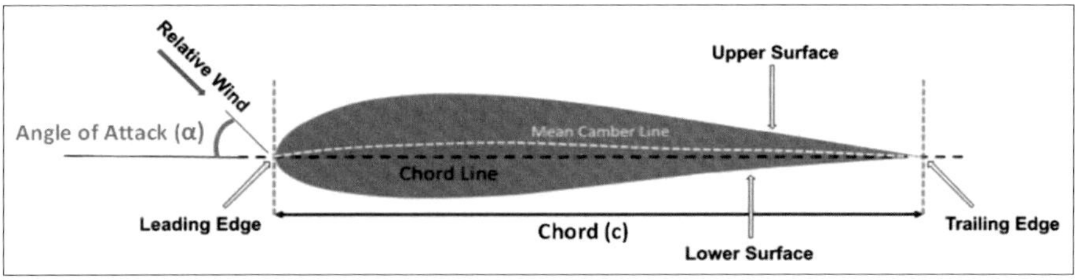

Figure 6.3 Characterization of a 2D airfoil.

An airfoil is for producing lift while minimizing drag. It is reported that airfoil performance is greatly influenced by the geometries when the upper surface is warmer than the ambient and exhibits varying degrees of lift improvement and drag reduction, see Table 6.1 [15]. Therefore, the choice of airfoil is crucial. A symmetrical 2-dimensional NACA0012 airfoil is selected in this study because of its simple geometry. The simple airfoil ensures any effects of warming the upper surface on the airfoil performance reliant on the geometry of the airfoil. It also eliminates the effects of maximum camber thickness or curves of more complex airfoils since for the NACA0012 airfoil, it is essentially zero.

The geometry of the NACA0012 airfoil was created in Gmsh using Eq. (6.4) [24]. In Eq. (6.4), the $\pm$ in the y-direction represents the symmetry of the airfoil along the x-axis. The independent x terms represent the extension of the airfoil in the x-direction (chord) and the dependent y term shows the extension of the airfoil in the y-direction (thickness), therefore, the 2D airfoil is confined in the x- and y-directions.

$$
\begin{aligned}
y = \pm\, & 0.594689181 \\
& \times [0.298222773 \times x^{0.5} - 0.127125232 \times x^{1} \\
& \quad - 0.357907906 \times x^{2} + 0.291984971 \times x^{3} \\
& \quad - 0.105174606 \times x^{4}]
\end{aligned} \tag{6.4}
$$

Using data from the NASA turbulence modeling resource, the airfoil is created between $x = 0$ and $x = 1.008930411365$ (the trailing edge is sharp at this location). Then the airfoil is scaled down by a factor of 1.008930411365 resulting in the replication of the 2D NACA0012 airfoil with $c = 1$ m. Scaling down the airfoil in OpenFOAM by 1/50th factor gives NACA0012 airfoil under study with $c = 20$ mm.

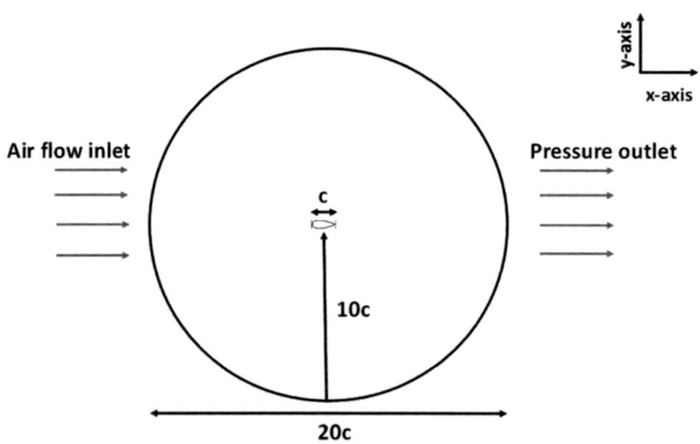

Figure 6.4 Dimensions of the computational domain in terms of chord length c.

Following the convention, the chord length (c) and thickness (δ) of the airfoil are taken as the characteristic lengths in the x- and y-directions, respectively. Dimensions of the computational domain as shown in Fig. 6.4 are scaled by the chord length c.

The radius of the circular domain is $10c$. The airfoil is located at approximately $10c$ downward of the uniform inlet stream. The reference of the coordinate system is fixed at the center of the airfoil, corresponding to ($10c$, $10c$). The velocity of the inlet stream of air in the x-direction is set at 24 m/s, and T_a = 300 K. This computational domain results in a turbulent flow with $Re = 3.3 \times 10^4$.

6.2.2 Governing Equations

The Navier-Stokes equation lies at the heart of all computational fluid dynamics simulations. Although the name Navier Stokes was initially restricted to the conservation of linear momentum, nowadays, the Navier-Stokes equation is used to denote the conservation of mass, momentum, and energy altogether. Due to the broad applicability of the equation ranging from the simulation of hurricanes to blood flow in an artery, it is essential to walk through the fundamentals.

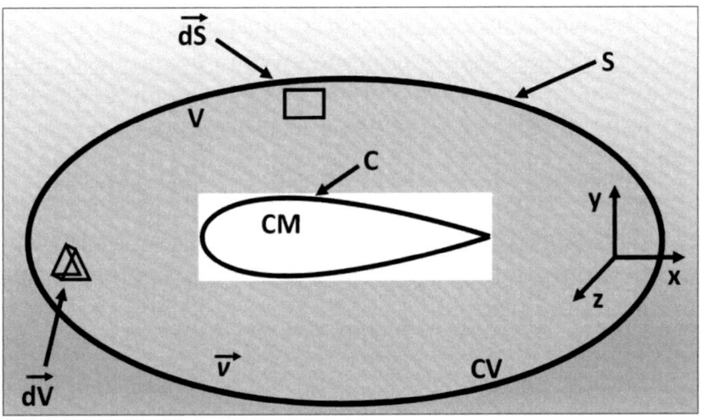

Figure 6.5 Airfoil control mass in a control volume and integration contour.

Consider an airfoil control mass enclosed in a control volume (CV) of volume (V), with control surface (S) in an inertial reference frame (x, y, z), as depicted in Fig. 6.5. The airfoil control mass has a constant mass m, density ρ, and velocity v. The flow field enclosed in the control volume around the airfoil varies both in space and time. It is evident that in the absence of any mass source or sink, a region will conserve its mass locally. Therefore, the mass will remain conserved. Reynolds Transport Theorem is a general conservation law that relates control mass conservation law to the control volume under consideration [25].

Using Reynolds Transport Theorem on the Eulerian Coordinate System (x, y, z), the expression for mass conservation is

$$\int_{CV}^{V}\left[\frac{D\rho}{Dt}+\rho\nabla\cdot\boldsymbol{v}\right]=0. \tag{6.5}$$

Since the expression in Eq. (6.5) is valid for an arbitrary control volume, the integral should indeed be equal to zero for it to be true for any geometry. Therefore, we can remove the integral, and the differential form of the mass conservation or continuity equation is obtained as Eq. (6.6).

$$\frac{D\rho}{Dt}+\rho\nabla\cdot\boldsymbol{v}=0 \tag{6.6}$$

Now for a material volume of a substance, Newton's second law of motion asserts that the momentum of a specific volume can change only in the presence of a net force acting on it, which could include both surface forces (f_s) and body forces (f_b). Therefore, considering the control volume in Fig. 6.5, its total momentum should remain conserved due to the absence of a net external force. Again, using Reynolds Transport Theorem on the control volume for momentum conservation, the expression for momentum conservation becomes,

$$\int_{CV}^{V}\left[\frac{\partial[\rho\boldsymbol{v}]}{\partial t}+\nabla\cdot\rho\boldsymbol{v}\boldsymbol{v}-\boldsymbol{f}\right]dV=0. \tag{6.7}$$

Here, $f = f_s + f_b$ and CV is an arbitrary geometry in space that contains control mass. For Eq. (6.7) to be true for any arbitrary geometry, the integral must indeed be equivalent to zero and the momentum equation in differential form is represented below,

$$\frac{\partial[\rho\boldsymbol{v}]}{\partial t}+\nabla\cdot\rho\boldsymbol{v}\boldsymbol{v}-\boldsymbol{f}=0 \tag{6.8}$$

It is worth mentioning that $\boldsymbol{f}_s$ and $\boldsymbol{f}_b$ terms can further be expanded to reveal a more comprehensive version of Eq. (6.8). However, such derivations are beyond the scope of the presented study, and further details can be found in existing literature [26, 27].

The energy equation is not coupled with the mass and momentum equation for incompressible flows. Instead, it is solved independently after obtaining velocity and pressure fields from the mass and momentum equations. However, for compressible flows, density and viscosity no longer remain constant, and their effects need to be accounted for with variations in temperature simultaneously. Consider the random motion of gas molecules in a closed container. As temperature increases, the gas molecules gain energy, and their random motion increases, leading to viscosity and density changes. To account for such changes, Sutherland's law is used in this work to relate the dynamic viscosity μ with temperature [28].

Furthermore, for a closed, isolated system, the sum of all forms of energy remains constant. In Fig. 6.5, the control volume has a total energy per unit mass of *e*. If $\dot{q}_v$ represents the rate of heat source/sink within the material volume per unit volume and $\dot{q}_s$ is the rate of heat transfer per unit area across the surface area of the material element, then invoking the First Law of Thermodynamics, the governing energy equation becomes,

$$\frac{\partial(\rho e)}{\partial t} + \nabla \cdot [\rho \boldsymbol{v} e] = -\nabla \cdot \dot{\boldsymbol{q}}_s - \nabla \cdot [p\boldsymbol{v}] + \nabla \cdot [\tau \cdot \boldsymbol{v}] + \boldsymbol{f}_b \cdot \boldsymbol{v} + \dot{q}_v \tag{6.9}$$

Here, *p* is the pressure distribution around the control volume. Equation (6.9) can be written in terms of specific internal energy, specific total enthalpy, or temperature.

It comes as no surprise that when temperature changes, the density of gas changes. The nature of density variation plays an important role in solving Eqs. (6.5–6.9). If the molecules involved are that of an ideal gas with pressure *p* and *R* is the gas constant, then the relationship between density and temperature can be represented using the ideal gas equation,

$$\rho = \frac{p}{RT} \tag{6.10}$$

While deriving the equation of motions, no mention was made as to whether the flow is laminar or turbulent. The laminar flow is stable and streamlined, whereas turbulent flow is random and more diffusive, causing rapid inter-mixing and formation of eddies. Due to the highly random fluctuations, a deterministic approach is not possible for turbulent flow solutions. Thus, Reynolds came forward with an averaging concept named Reynolds Averaged Navier Stokes (RANS). Through the concept of RANS, all flow quantities are expressed as the sum of mean and fluctuating components. Therefore, in all the equations mentioned above, instead of just considering steady flow quantities, their time-averaged terms must be substituted (consisting of mean and fluctuating components) to account for the turbulence generated. The step is crucial since the simulations performed in this study at $Re = 3.3 \times 10^4$ are under turbulent flow conditions.

6.2.3 Simulation, Verification, and Validation

The system closure was realized using the RANS Spalart-Allmaras [29] model. The one-equation Spalart-Allmaras model is used because, unlike other algebraic models [30, 31] and one-equation models [32–34], the Spalart-Allmaras model is local. The locality means that the model is compatible with grids of any structure (structured and unstructured) and Navier-Stokes solvers in 2D and 3D. Then the system of steady-state compressible Navier-Stokes equations is solved using rhoSimpleFoam solver, and the variable fluid properties are simulated using the thermophysical library of OpenFOAM.

The mesh around the geometry of the 2D NACA0012 airfoil is generated using Gmsh and converted to an OpenFOAM environment. The entire domain mesh and zoom-in of the finer mesh near the airfoil are illustrated in Fig. 6.6. The airfoil is then split into the upper and lower surface patches to implement temperatures warmer and cooler than the ambient.

To begin, mesh independence and convergence studies are carried out at $Re = 3.3 \times 10^4$ with no surface heating or cooling at a Mach number (Ma) of 0.0725 and $T_a = 300$ K. Figure 6.7a shows the variation of C_L at four AOA values and its convergence with mesh refinement. While using a fine mesh, with 15 × 10^4 nodes and beyond, no significant change in the solution is obtained at all AOA values. However, for meshes with less than 2.5 × 10^4 nodes, the solution diverges. Figure 6.7b shows the variation of C_D at the four studied AOA values. A coarse mesh overpredicts the value of the drag coefficient significantly. The overprediction decreases rapidly with an increasing number of elements, and the asymptotic value is reached when the number of nodes is 15 × 10^4. Thus, a mesh consisting of 15 × 10^4 nodes, 130970 quadrangles, and 35364 triangles was chosen for all the further simulations performed.

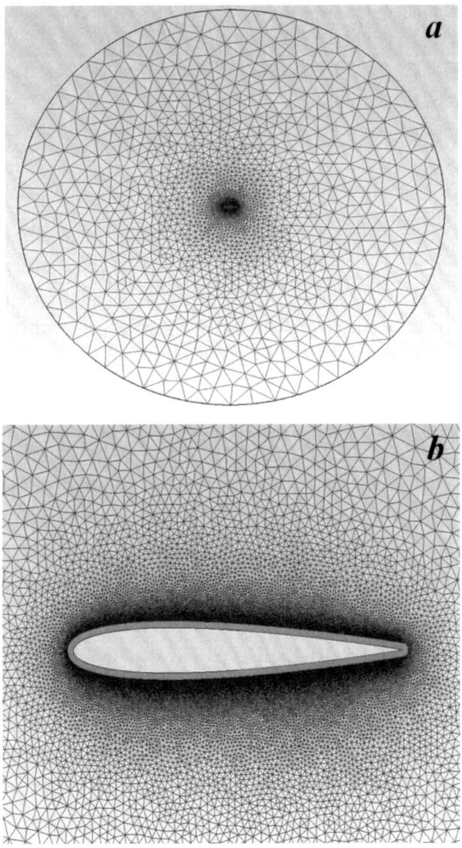

Figure 6.6 Domain mesh around the 2D airfoil: (a) whole domain and the refinement boxes and (b) zoom-in at the airfoil surroundings.

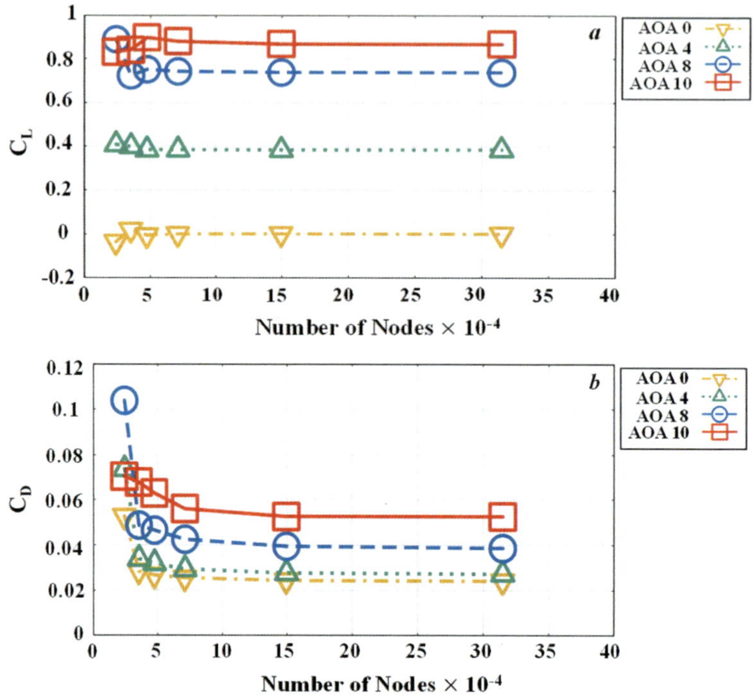

Figure 6.7 Grid independency results: The effects of number of nodes on (a) C_L and (b) C_D are obtained at Ma = 0.0725, $Re = 3.3 \times 10^4$, and T_a = 300 K.

To validate the code, an airflow with Ma = 0.0725, at a pressure of 1 atm, and T_a = 300 K is set up in the airfoil domain and simulated. The airfoil's upper and lower surfaces are maintained at a temperature of 300 K and the flow $Re = 3.3 \times 10^4$.

To facilitate a generalized expression for easier interpretation, the temperature difference is normalized as:

$$\frac{\Delta T}{T_a} = \frac{(\text{Upper surface temperature} - \text{Lower surface temperature})}{\text{Ambient temperature}} \tag{6.11}$$

Figure 6.8a shows the variation of C_L with AOA at $Re = 3.3 \times 10^4$, Ma = 0.0725, and T_a = 300 K. The upper surface of the airfoil is held at a constant temperature of 300 K. The corresponding C_L/C_D ratio is plotted as a function of AOA, as shown in Fig. 6.8b, and the results are validated with data obtained by

Kim et al. [19]. The presented results are in good agreement. While both studies employ the Spalart-Allmaras model for simulating turbulence, the grid size and shape are different. In our study, a circular control surface is used; whereas, a parabolic control surface was deployed by Kim et al. [19]. Furthermore, the present simulations are performed in OpenFOAM, while Kim et al. [19] utilized FLUENT.

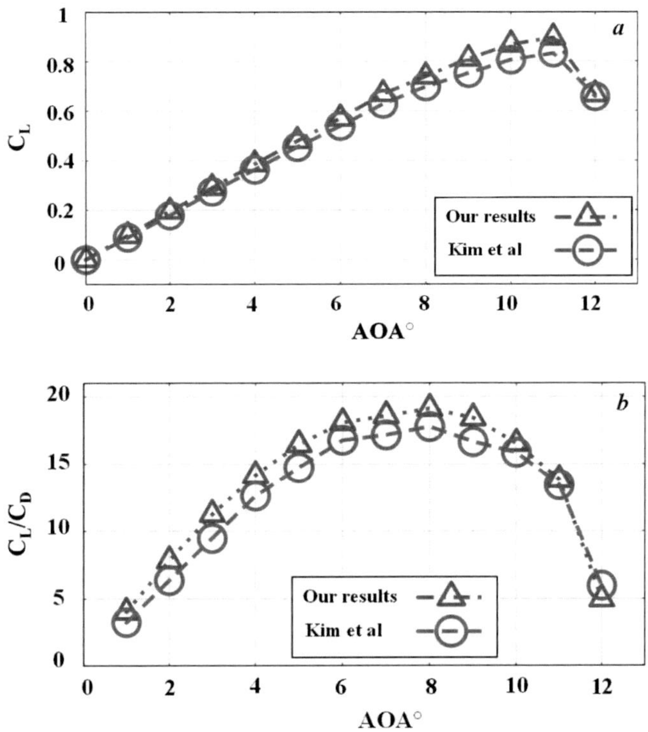

Figure 6.8 Validation of present work: The effects of AOA on (a) C_L and (b) C_L/C_D of NACA0012 airfoil at Ma = 0.0725, Re = 3.3 × 10^4, and T_a = 300 K.

6.3 Results and Discussion

Previous research [14, 15] has shown that the black-colored wings of birds, such as the albatross, warm up faster than the white-colored ones. The differing solar absorptivity of black and white colors causes uneven warming. The uneven warming sets up a temperature difference between the top and bottom surfaces of the wings. Their study revealed a temperature difference of almost 30 °C between the top black and white bottom wing. Thus, a temperature difference of 30 K is used in the present study to investigate warmer and cooler than the ambient upper surface on the performance of the NACA0012 airfoil.

The upper surface of the NACA0012 airfoil, at T_a = 300 K, Re = 3.3 × 10^4, and Ma = 0.0725, is warmed to a uniform temperature of 330 K (ΔT = 30 K and $\frac{\Delta T}{T_a}$ = 0.1). The surface temperatures being fixed, we do not consider heat transfer in the solid airfoil via conduction (from the hot to the cold surface). Convective heat transfer, as shown in Fig. 6.9, is modeled using the rhoSimpleFoam solver and thermophysical library of OpenFOAM. The force coefficients are calculated when the upper surface temperature is the same and warmer than the ambient. The results, presented in Fig. 6.10, reveal that C_L decreases, whereas C_D increases when the upper surface is warmer for all studied AOA values. The effect of a warmer upper surface is more pronounced at higher AOA values, especially when AOA = 11°. A decrease in C_L value from 0.896 to 0.860 and an increase in C_D value from 0.065 to 0.070 was obtained. It corresponds to a drop in C_L/C_D value from 13.79 to 12.35, as shown in Fig. 6.11.

Figure 6.9 Convective heat transfer from the warm upper surface of the airfoil (T_u = 330 K) to the ambient (T_a = 300 K).

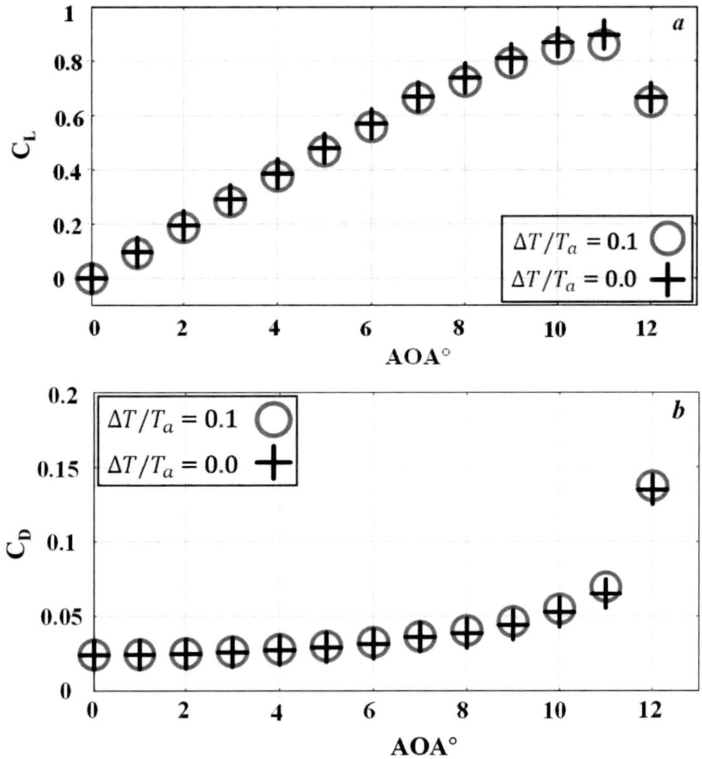

Figure 6.10 The effects of heating on the lift and drag at *Re* = 3.3 × 10^4 and *Ma* = 0.0725: (a) C_L vs AOA and (b) C_D vs AOA.

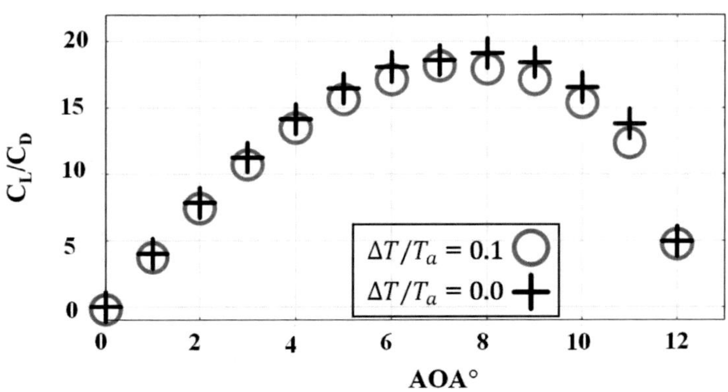

Figure 6.11 Effect of heating upper surface on C_L/C_D vs AOA at *Re* = 3.3 × 10^4, $\frac{\Delta T}{T_a}$ = 0.1, $\frac{\Delta T}{T_a}$ = 0.0, and *Ma* = 0.0725.

To uncover the underlying physics behind the observed changes in C_L and C_D, the pressure and wall-shear stress values on the airfoil's upper surface are plotted against the geometry of the airfoil to visualize the distribution. The pressure and wall-shear stress are plotted for the largest observed changes in lift and drag at AOA = 11°.

As shown in Fig. 6.12, the value of pressure for the $\frac{\Delta T}{T_a} = 0.1$ case is larger when compared to the reference $\frac{\Delta T}{T_a} = 0$ case. The increase is pronounced near the leading edge of the airfoil, with the maximum increase occurring at approximately $x/c = 0.25$ (the leading edge is at $x/c = 0$ and $c = 20$ mm). The pressure difference reduces as we move along the airfoil's chord toward the trailing edge, and it attains almost the same pressure as the airfoil's upper surface for the no-heating case.

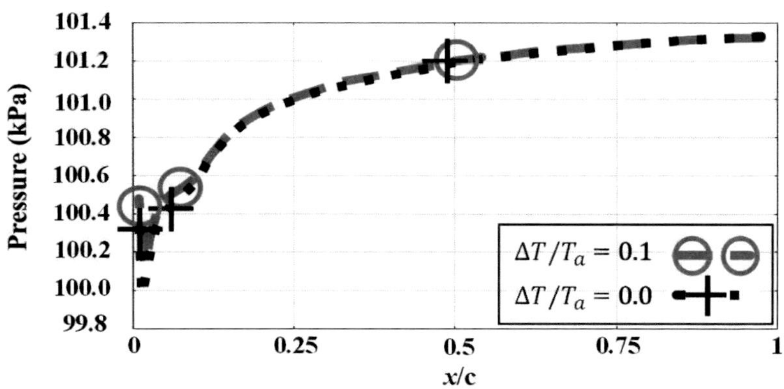

Figure 6.12 Pressure (kPa) along the upper surface of the airfoil at AOA = 11°.

Furthermore, a similar trend is observed by the wall shear stress as in the case of pressure, as shown in Fig. 6.13. A significant change between the warmer and normal upper surfaces of the airfoil peaks around the leading edge of the airfoil. The difference is reduced as we move toward the airfoil's trailing edge. A noteworthy feature here is that the wall shear stress increases from 0.83 Pa to 1.40 Pa near the airfoil's leading edge. The exact location corresponds to approximately $x/c = 0.25$. The present results show that wall shear stress at the upper surface of the airfoil increases with warmer temperatures than ambient. The reason is that for air, both the dynamic and kinematic viscosity increase with temperature. As a result, the shear increases, which slows down the fluid flow and increases the pressure.

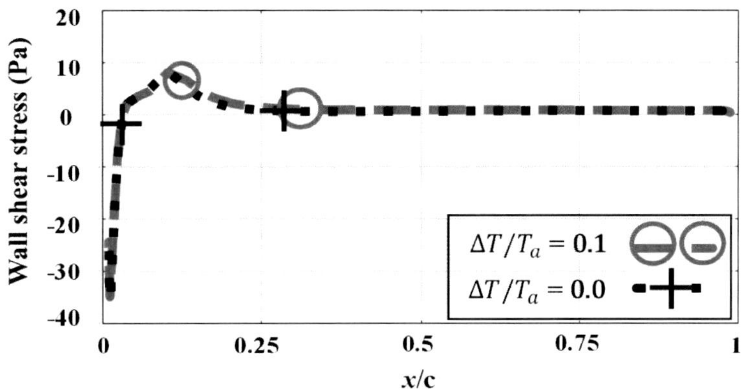

Figure 6.13 Wall shear stress (Pa) along the upper surface of the airfoil at AOA = 11°.

The present simulations suggest that upper surface heating of the NACA0012 airfoil causes a rise in drag and a reduction in the lift by increasing the pressure mainly around the nose region of the airfoil. The increased pressure pushes the airfoil downwards and results in reduced lift. On the other hand, the increased value of wall shear stress in the x-direction (the airflow direction) and the increased viscosity negatively affect the drag.

The results of our study contradict the concept of birds utilizing surface heating to optimize their flight efficiency. Since the warmer upper surface shows a negative effect, considering a cooler upper surface seems the next appropriate choice. For this purpose, the upper surface of the NACA0012 airfoil was cooled to 270 K (ΔT = –30 K and $\frac{\Delta T}{T_a}$ = -0.1) and the simulations were repeated under the same flow conditions. Like the case of warming, for cooling, heat transfer from the cooler airfoil surface to the warmer one via conduction in the solid airfoil is not taken into consideration. Convective heat transfer, as shown in Fig. 6.14, is modeled similarly using the rhoSimpleFoam solver and thermophysical library of OpenFOAM. The effect of a cooler upper surface on the values of C_L and C_D is presented in Fig. 6.15.

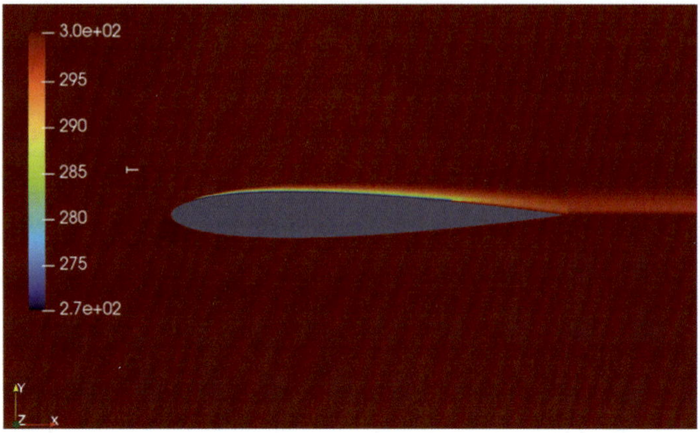

Figure 6.14 Convective heat transfer from the cool upper surface of the airfoil (T_u = 270 K) to the ambient (T_a = 300 K).

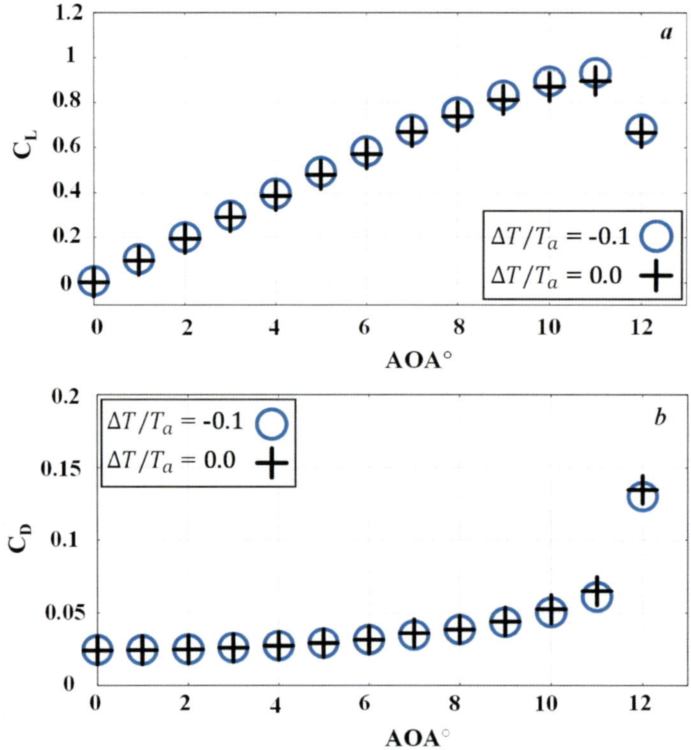

Figure 6.15 The effect of cooling on lift and drag: (a) comparison of C_L vs AOA and (b) comparison of C_D vs AOA.

The lift increases, while the drag reduces, and the maximum increase in C_L/C_D occurs at AOA = 11°. The C_L increases from 0.896 to 0.933, whereas C_D decreases from 0.065 to 0.061. The corresponding value of C_L/C_D increases from 13.8 to 15.3, as shown in Fig. 6.16.

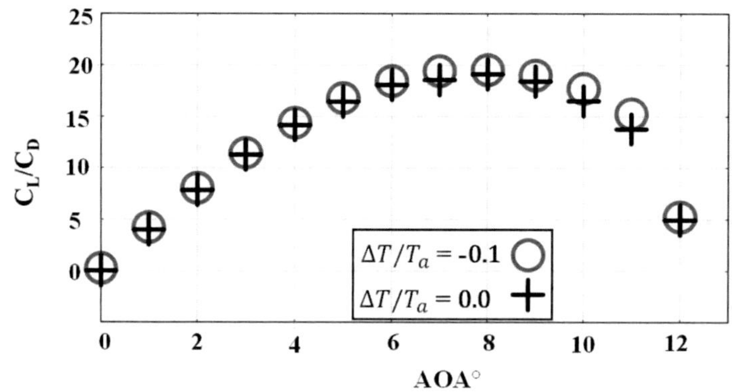

Figure 6.16 The effect of cooling on C_L/C_D vs AOA.

As shown in Fig. 6.17, the value of pressure for the $\frac{\Delta T}{T_a}$ = -0.1 case is lower compared to the reference $\frac{\Delta T}{T_a}$ = 0 case. It is the front end of the airfoil that becomes the most susceptible to change with a cooler upper surface. As we move toward the trailing edge, the pressure evens out and becomes almost like the reference $\frac{\Delta T}{T_a}$ = 0 case.

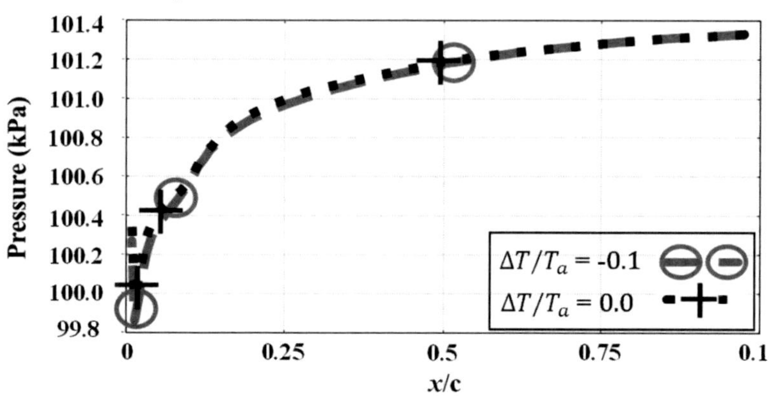

Figure 6.17 Pressure (kPa) along the upper surface of the airfoil at AOA = 11°.

The wall shear stress in the x-direction decreases when the upper surface is cooled; compare $\frac{\Delta T}{T_a}$ = −0.1 case with the reference $\frac{\Delta T}{T_a}$ = 0.0 case, shown in Fig. 6.18. The value decreases from 0.83 Pa to 0.35 Pa at x/c = 0.25. The decrease starts close to the airfoil's leading edge and as we move toward the trailing edge, the deviation of wall shear from $\frac{\Delta T}{T_a}$ = 0.0 case becomes negligible.

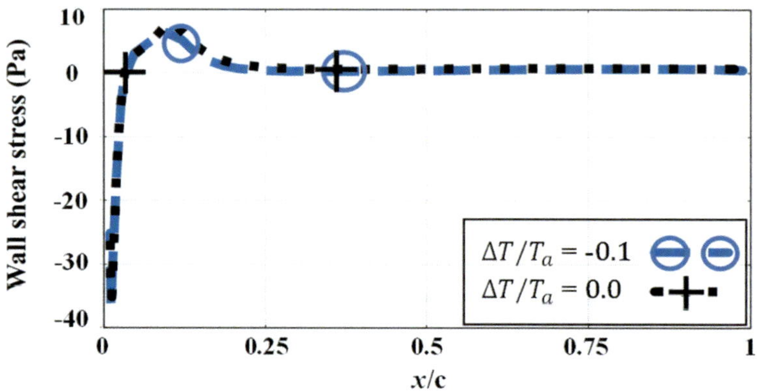

Figure 6.18 Wall shear stress (Pa) along the upper surface of the airfoil at AOA = 11°.

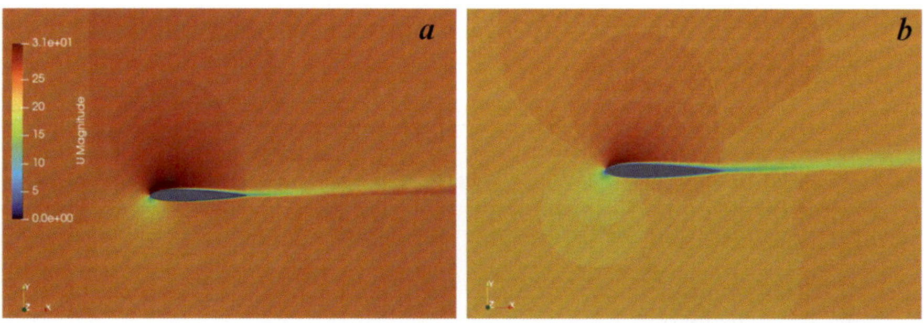

Figure 6.19 Velocity distribution around the airfoil: (a) $\frac{\Delta T}{T_a}$ = −0.1 and (b) $\frac{\Delta T}{T_a}$ = 0.1.

The reduced pressure creates the necessary pressure difference near the airfoil's leading edge to increase the lift coefficient in the y-direction. At the same time, the reduced wall shear stress in the x-direction reduces the drag coefficient. The reason is that with decreasing temperature, the viscosity decreases. As a result, the local velocity near the leading edge of the airfoil increases, causing the pressure to decrease. Figure 6.19 highlights the increase in velocity on cooling near the leading edge when compared to the warmer case. A combination of these factors makes upper surface cooling a better alternative to improve the flight dynamics of the NACA0012 airfoil.

6.4 Conclusions

A study conducted in Ref. [14] found that upper surface heating improved aerodynamic performance. Similar tests on a NACA0012 airfoil, however, have shown that upper surface cooling improves performance [17–20]. The current investigation stems from the previously mentioned literature contrast. The effects of temperatures warmer and cooler than the ambient on the aerodynamic performance of a simple 2D NACA0012 airfoil were explored.

To understand the effects of such thermal camber, 2D numerical simulations were performed using OpenFOAM to understand the aerodynamic performance of NACA0012 airfoil subjected to turbulent flow conditions ($Re = 3.3 \times 10^4$) at $T_a = 300$ K, $T_u = 300 \pm 30$ K, $p = 1$ atm, and $Ma = 0.0725$. The major findings of the study are summarized as follows:

When the upper surface of the NACA0012 airfoil is kept at a uniform temperature warmer than the ambient ($\frac{\Delta T}{T_a} = 0.1$), the aerodynamic performance of the airfoil suffers negatively. The C_L value decreases from 0.896 to 0.860 (4% reduction in lift) and the C_D value increases from 0.065 to 0.070 (8% increase in drag). Overall, the maximum magnitude of reduced performance was observed at AOA 11° with the C_L/C_D value decreasing from 13.79 to 12.35 (10% drop in performance).

Since warmer temperatures showed a negative effect, the upper surface of the airfoil was further tested at a uniform temperature cooler than the ambient ($\frac{\Delta T}{T_a} = -0.1$) and it improved the aerodynamic performance. The C_L value increased from

0.896 to 0.933 (4% rise in lift) and the C_D value decreased from 0.065 to 0.061 (6% reduction in drag). Again, the maximum enhancement in performance was observed at AOA 11° with the C_L/C_D value increasing from 13.8 to 15.3 (11% improvement).

The reported 11% increase or 10% decrease has been obtained at an AOA of 11° at upper surface temperate of 270K and 300K, respectively, with T_a = 300 K, p = 1 atm, and Ma = 0.0725.

Contrary to the existing bioinspired studies on upper surface heating of airfoils, our study finds that a cooler upper surface of the NACA0012 airfoil than the ambient improves the aerodynamic performance. The improvement occurs due to the reduced pressure and wall shear stress that increase the lift and reduce the drag forces, respectively.

It appears that Ref. [14] is contrary to other studies, which is an anomaly. It is important to note that the wings of birds are three-dimensional and have intelligently designed morphology, whereas a simple, smooth, two-dimensional airfoil has been simulated in studies such as ours. It should be noted that although the NACA0012 airfoil does not mimic the wings of birds exactly, the present study lays the foundation for analyzing thermal camber in airfoils inspired by birds like albatross.

Further modifications can be added to the geometry of the NACA0012 airfoil to resemble in much more depth, the actual wings of birds. 3D simulations form the next step of the presented study, to capture the vortices around the airfoil and its change with further bioinspired studies.

Acknowledgments

This work is made possible by MITACS Globalink Research Internship Program Canada. The lead author would like to thank MITACS for providing the enriching opportunity to conduct research at the University of Windsor. Some of the simulations presented in this paper were carried out using the Param Utkarsh (PU) HPC facility of the National Supercomputing Mission (NSM), Government of India.

References

1. Shakeri, R., Al-Garadi, M. A., Badawy, A., Mohamed, A., Khattab, T., Al-Ali, A. K., Harras, K. A., Guizani, M. (2019). Design Challenges of Multi-UAV Systems in Cyber-Physical Applications: A Comprehensive Survey and Future Directions. *IEEE Communications Surveys & Tutorials*, 21(4), 3340–3385.
2. Shakhatreh, H., Sawalmeh, A., Al-Fuqaha, A., Dou, Z., Almaita, E., Khalil, I., Othman, N. S., Khreishah, A., Guizani, M. (2018). Unmanned Aerial Vehicles: A Survey on Civil Applications and Key Research Challenges. *IEEE Access*, 7, 48572–48634.
3. Balestrieri, E., Daponte, P., De Vito, L., Lamonaca, F. (2021). Sensors and Measurements for Unmanned Systems: An Overview. *In Sensors*, 21(4), 1–27.
4. Yu, J. (2018). Design and Optimization of Wing Structure for a Fixed-Wing Unmanned Aerial Vehicle (UAV). *Modern Mechanical Engineering*, 08(04), 249–263.
5. Sultan, L., Anjum, M., Rehman, M., Murawwat, S., Kosar, H. (2021). Communication among Heterogeneous Unmanned Aerial Vehicles (UAVs): Classification, Trends, and Analysis. *IEEE Access*, 9, 118815–118836.
6. Townsend, A., Jiya, I. N., Martinson, C., Bessarabov, D., Gouws, R. (2020). A Comprehensive Review of Energy Sources for Unmanned Aerial Vehicles, Their Shortfalls, and Opportunities for Improvements. *Heliyon*, 6(11). Elsevier Ltd.
7. Coyle, E. D., Simmons, R. A. (2014). *Understanding the Global Energy Crisis.* Purdue University Press Books, West Lafayette, 2014-03-01.
8. Bushnell, D. M., Moore, K. J. (1991). Drag Reduction in Nature. *Annu. Rev. Fluid Mech*, 23, 65–79.
9. Tucker, V. A. (1993). Gliding Birds: Reduction of Induced Drag by Wing Tip Slots Between the Primary Feathers. *In J. exp. Biol*, 180, 285–310.
10. Floreano, D., Wood, R. J. (2015). Science, Technology, and the Future of Small Autonomous Drones. *Nature*, 521(7553), 460–466. Nature Publishing Group.
11. Richardson, P. L., Wakefield, E. D., Phillips, R. A. (2018). Flight Speed and Performance of the Wandering Albatross with Respect to Wind. *Movement Ecology*, 6(1).
12. Hedenström, A. (2002). Aerodynamics, Evolution, and Ecology of Avian Flight. *Trends in Ecology & Evolution*, 17(9), 415–422.

13. Hassanalian, M., Abdelmoula, H., Ayed, S. ben, Abdelkefi, A. (2017). Thermal Impact of Migrating Birds' Wing Color on Their Flight Performance: Possibility of New Generation of Biologically Inspired Drones. *Journal of Thermal Biology*, 66, 27–32.
14. Hassanalian, M., Ayed, S. ben, Ali, M., Houde, P., Hocut, C., Abdelkefi, A. (2018). Insights on the Thermal Impacts of Wing Colorization of Migrating Birds on Their Skin Friction Drag and the Choice of Their Flight Route. *Journal of Thermal Biology*, 72, 81–93.
15. Hassanalian, M., Pellerito, V., Sedaghat, A., Sabri, F., Borvayeh, L., Sadeghi, S. (2019). Aerodynamics Loads Variations of Wings with Novel Heating of Top Surface: Bioinspiration and Experimental Study. *Experimental Thermal and Fluid Science*, 109.
16. Rogalla, S., D'Alba, L., Verdoodt, A., Shawkey, M. D. (2019). Hot Wings: Thermal Impacts of Wing Coloration on Surface Temperature During Bird Flight. *Journal of the Royal Society Interface*, 16(156).
17. Norton, D. J., Macha, J. M., Young, J. C. (1973). Surface Temperature Effect on Subsonic Stall. *Journal of Spacecraft and Rockets,* 10(9), 581–587.
18. Hinz, D. F., Alighanbari, H., Breitsamter, C. (2013). Influence of Heat Transfer on the Aerodynamic Performance of a Plunging and Pitching NACA0012 Airfoil at Low Reynolds Numbers. *Journal of Fluids and Structures*, 37, 88–99.
19. Kim, J., Rusak, Z., Koratkar, N. (2003). Small-Scale Airfoil Aerodynamic Efficiency Improvement by Surface Temperature and Heat Transfer. *AIAA Journal*, 41(11), 2105–2113.
20. Bekka, N., Bessaïh, R., Sellam, M., Chpoun, A. (2009). Numerical Study of Heat Transfer Around the Small-Scale Airfoil Using Various Turbulence Models. *Numerical Heat Transfer; Part A: Applications,* 56(12), 946–969.
21. Samiee, A., Djavareshkian, M. H., Feshalami, B. F., Esmaeilifar, E. (2018). Improvement of Airfoils Aerodynamic Efficiency by Thermal Camber Phenomenon at Low Reynolds Number. *Journal of Aerospace Technology and Management*, 10.
22. Longo, J. M. A., Radespiel, R. (1995). Numerical Simulation of Heat Transfer Effects on 2-D Steady Subsonic Flows. *26th AIAA Fluid Dynamics Conference,* AIAA 95–2298.
23. Ribeiro, D. G., Bravo-Mosquera, P. D., Muñoz, H. D. C. (2019). Heat Transfer Effects on Aerodynamic Performance of a S809 Airfoil for Wind Turbine Application. AIAA Propulsion and Energy Forum and Exposition, 19–22 August 2019.

24. NACA 0012 2D Airfoil Validation. From https://turbmodels.larc.nasa.gov/naca0012_val.html.
25. Gatski, T. B., Bonnet, J.-P. (2009). Kinematics, Thermodynamics, and Fluid Transport Properties. *Compressibility, Turbulence and High-Speed Flow*, pp. 1–19.
26. Anderson, J. (2017). *Fundamentals of Aerodynamics*. McGraw-Hill Education, 2 Penn Plaza, New York.
27. Moukalled, F., Mangani, L., Darwish, M. (2016). Fluid Mechanics and Its Applications. *The Finite Volume Method in Computational Fluid Dynamics*. (Vol. 113, Springer International Publishing Switzerland).
28. Sutherland, W. (1893). LII. The Viscosity of Gases and Molecular Force. *The London, Edinburgh, and Dublin Philosophical Magazine and Journal of Science*, 36(223), 507–531.
29. Allmaras, S. R. (1992). A One-Equation Turbulence Model for Aerodynamic Flows. AIAA-92-G439.
30. Baldwin, B. S., Lomax, H., (1978). Thin Layer Approximation and Algebraic Model for Separated Turbulent Flows. *AIAA 16th Aerospace Sciences Meeting* (16–18 January, Huntsville, AL, U.S.A).
31. Johnson, D. A., King, L. S. (1985). A Mathematically Simple Turbulence Closure Model for Attached and Separated Turbulent Boundary Layers. *AIAA Journal*, 23(11), 1684–1692.
32. Bradshaw, P., Ferriss, D. H., Atwell, N. P. (1967). Calculation of Boundary-Layer Development Using the Turbulent Energy Equation. *Journal of Fluid Mechanics*, 28(3), 593–616.
33. Nee, V. W., Kovasznay, L. S. G. (1969). Simple Phenomenological Theory of Turbulent Shear Flows. *Physics of Fluids*, 12(3), 473–484.
34. Thomas, J. L., Salas, M. D. (1986). Far-Field Boundary Conditions for Transonic Lifting Solutions to the Euler Equations. *AIAA Journal*, 24(7), 1074–1080.

Chapter 7

Thermal Regulation due to Vertical Greening Systems: A Review

Saumili Jana,[a,b] Rajaram Lakkaraju,[a] Jacqueline A. Stagner,[b] and David S.-K. Ting[b]

[a]*Department of Mechanical Engineering, Indian Institute of Technology Kharagpur, India*
[b]*Turbulence & Energy Lab, University of Windsor, Canada*

rlakkaraju@mech.iitkgp.ac.in

Green infrastructure is increasingly becoming popular in cities to counter the negative effects of urbanization, decreasing green cover, and climate change. Vertical greening systems (VGSs) are one such green infrastructure and are gaining recognition as they lead to effective utilization of vertical spaces to increase the green cover in urban areas. These not only help increase the aesthetics of the neighborhood and provide general benefits to society but also result in passive thermal regulation of a building and its surroundings, leading to building energy savings. In this study, a review of the existing literature on vertical greening systems, with a focus on thermal performance and building energy transfer, is provided. The system is simplified and explained mathematically, which is essential when optimizing the design for maximum building energy savings in the long run. It is difficult to compare

Reclaiming Eden: Responsible Living, Engineering, and Architectures
Edited by David S.-K. Ting and Jacqueline A. Stagner

ISBN 978-981-5129-31-1 (Hardcover), 978-1-003-49612-0 (eBook)
www.jennystanford.com

the results of different studies due to the differences in study parameters and the dynamic nature of the system. If VGSs are to be used as a future solution and implemented on large scales, it is necessary to devise a methodology to standardize the system and define universal performance indicators to compare different VGSs. This work proposes such indicators and provides a directional approach for future studies.

7.1 Introduction

The negative consequences of deforestation, industrialization, and urbanization are a global concern. Although more and more trees can be planted, eventually that could lead to a space crunch since land is a finite resource. Therefore, one needs to think of an innovative solution. Vertical greening systems could be one of the best approaches to resolve this issue and utilize limited space efficiently to increase the green cover in cities.

A VGS is a greening practice that involves growing plants on vertical surfaces such as buildings, walls, facades, fences, etc. It can help reduce the negative effects of decreasing green areas and increasing asphalt and concrete cover. The benefits of VGSs include potentially reducing air pollution, counterbalancing the urban heat island (UHI) effect, reducing noise, and increasing biodiversity, to name a few. VGSs also provide several other environmental, health, societal, and aesthetic benefits (Ghazalli et al., 2019; Medl et al., 2017). Most importantly, VGSs help in the passive thermal regulation of a building and its surroundings. In hot weather, it leads to a cooling effect inside the building, and, in cold weather, it leads to warm interiors (Medl et al., 2017). This thermal moderation due to a VGS leads to potential building energy savings in all seasons (Raji et al., 2015).

VGSs are not a novel invention but have a very rich and ancient history. The origins of incorporating vegetation into building design can be traced back to ancient civilizations, particularly in connection with ziggurats or stone structures. The Babylonian hanging gardens are widely regarded as the earliest documented instance of incorporating plant life onto buildings. These gardens likely consisted of terraces or rooftop gardens situated on ground-level structures, with vegetation cascading over the edges of the terraces (Hoffmann et al., 2023). Around 2000 years ago, vines were used as one of the earliest forms of building greening used for covering architectural structures in Mediterranean regions such as Rome to provide shading, cooling, and agricultural products (Hoffmann et al., 2023; Köhler, 2008). Woody vines, fruit espaliers, and ornamental flowering plants were commonly used to cover castles and houses in Central Europe around five hundred years ago (Köhler, 2008). The early 20th century saw climbers like English Ivy and Boston Ivy become popular for building and façade greening in Europe (Hoffmann et al., 2023; Köhler, 2008; Medl et al., 2017). In the 1980s, Bartfelder and Kohler initiated preliminary research endeavors focused on the ecological functions of façade greening, thereby establishing the foundation for this innovative research area (Medl et al., 2017). In the late 20th century, with rising concerns regarding environmental degradation, there was an increase in the tendency to incorporate more and more sustainable principles into architectural design. The first green wall using the hydroponic technology known as "Mur Vegetal" was built in 1994 and pioneered by the French botanist Patrick Blanc (Medl et al., 2017). Cities such as Berlin, Vienna, Ljubljana, and Taipei have shown enthusiasm for VGSs lately, and successfully completed several large-scale projects (Hoffmann et al., 2023). With the development of recent technologies and newer gardening techniques, different vertical greening methods are being developed. With advancements like hydroponics and the development of wall-based systems, climbing vegetation is no longer the key requirement (Hoffmann et al., 2023). Latest studies have been focusing on the possibility of simultaneously using green facades and photovoltaic systems on a building façade, since the green buffer layer provides a cooling effect and lowers the operating temperature of the photovoltaic module (Penaranda Moren & Korjenic, 2017; Wu & Chul-Soo, 2023). Given their advantages, VGSs are experiencing an ongoing expansion both in terms of popularity and advancements in technology.

Vertical greening systems are usually of interest when it comes to building thermal regulation and energy savings. However, VGSs are very complex systems to understand and model. Many studies have been conducted in order to understand the impact of the vegetation layer on building thermal regulation. This becomes important in the prediction of building energy requirements and designing the best possible VGS to maximize the thermal benefits and energy savings. In this study, the existing literature is reviewed to provide a clear picture of the existing knowledge on this topic. In the process, gaps in current knowledge are identified and possible solutions or corrective measures are suggested. Further, indicators for comparing the performance of different VGSs in different locations and under different conditions are proposed.

7.2 Classification of VGS

Vertical greening systems can be classified into different categories based on their features and architecture. Figure 7.1 provides a glimpse into the classification of VGSs. Broadly, VGSs can be divided into "green facades" and "green walls" depending on how plants are rooted (Köhler, 2008; Medl et al., 2017).

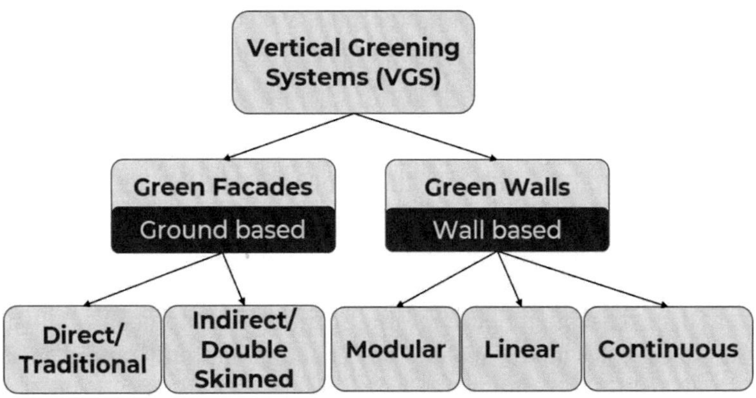

Figure 7.1 Classification of VGS.

The VGS is called a green façade if plants are rooted in the ground, as shown in Fig. 7.2. If the plants have no dependence on the ground, the VGS is known as a green wall. Green facades can be further classified into direct and indirect green facades; whereas, green walls can be further classified into modular, continuous, and linear green walls.

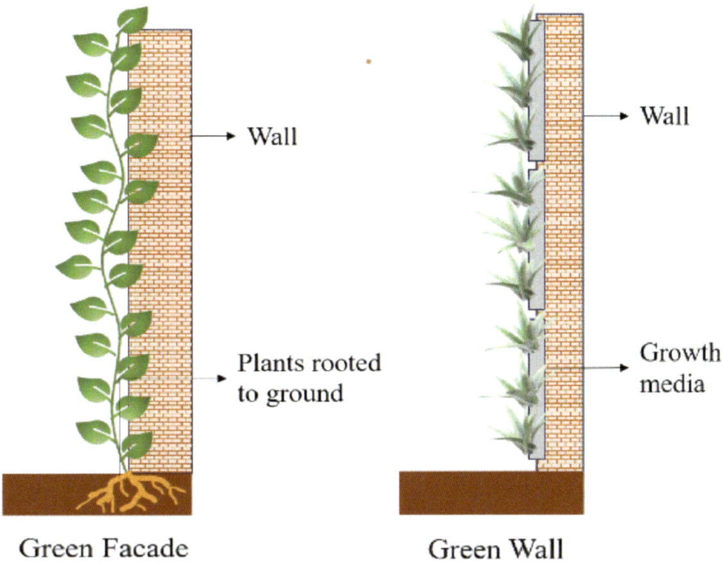

Figure 7.2 Illustration demonstrating difference between green façades and green walls.

7.2.1 Green Facades

Green facades are a ground-based greening method in which plants are grown directly in the soil, as depicted in Fig. 7.2. The plants used are usually climbers that grow upward and cover the vertical surfaces of the building. Green facades are further classified into "direct" and "indirect" green facades depending on the need for additional external supporting structures to support the upward growth of the plants. Figure 7.3 portrays the difference between the two major classifications of green facades.

The direct green facade is the simplest among all VGSs and has the most traditional architecture. In this case, plants do not require any additional supporting elements to aid their growth. The plants have adventitious clinging roots or self-adhesive pads, which help them hold on to the walls directly and climb upwards, as noted in Fig. 7.3(a). Plants like Boston Ivy (*Parthenocissus tricuspidata*) or English Ivy (*Hedera helix*) are well-suited for this type of green facade.

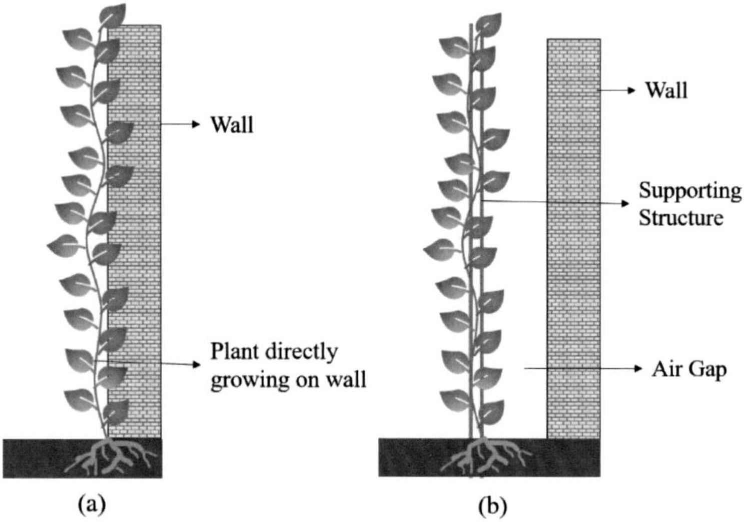

Figure 7.3 Representative image of (a) direct green façade and (b) indirect green façade.

In the case of indirect green facades, also known as double-skinned green facades, additional external supporting structures are required to assist the upward growth of the plants. These supporting elements include meshes, trellis, cables, or nets made up of stainless or galvanized steel, hardwood, plastics, etc. This usually leads to the formation of a second layer containing the vegetation at some distance from the walls, thus creating an air gap between the wall and the vegetation layer, as depicted in Fig. 7.3(b). These support structures also help increase the resistance of the system to environmental action such as rain, wind, etc. Plants with tendrils to climb upwards, such as grape vines, are ideal for this type of green facade (Hoffmann et al., 2023; Köhler, 2008; Palermo & Turco, 2020).

7.2.2 Green Walls

Green walls, also frequently referred to as living wall systems (LWSs), are a wall-based greening method. Green walls are a more recent innovation as compared to green facades and this method can lead to faster coverage by vegetation, especially in the case of tall buildings. They are usually designed with pre-vegetated panels or vertical modules fixed to the wall surface, and there is no dependence on the plants on the ground for rooting space or irrigation. Green walls can support a wide variety of plant species such as grasses, shrubs, perennial plants, succulents, etc., which can be selected based on factors such as climatic conditions or aesthetics. Green walls can be further classified as mainly "modular" or "continuous" green walls, as demonstrated in Fig. 7.4.

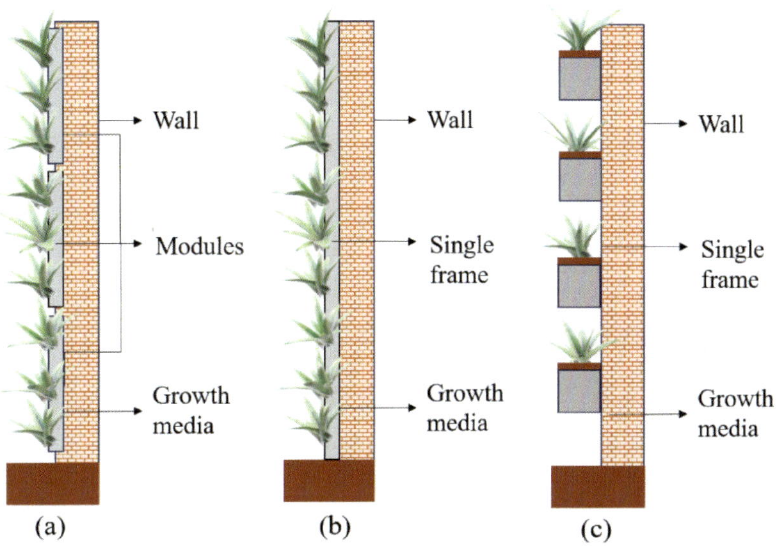

Figure 7.4 Representative image of (a) modular green wall, (b) continuous green wall, and (c) linear green wall.

In modular green walls, several modular elements such as pocket-type planters or panels are installed, which cover the entire vertical surface, as illustrated in Fig. 7.4(a). These modules hold the substrate and are attached to a fixed rigid support behind it. The growth media may be organic or inorganic. The irrigation system depends on the layout of the modules/panels and is usually installed in between the panels. The water is drained through the panels and is collected at the bottom of the entire green wall. Modular green walls allow greater seeding depth as compared to continuous green walls and are usually easier to maintain in terms of replacing damaged plants or changing plant species.

In contrast, continuous green walls consist of a single supporting structure/frame, as depicted in Fig. 7.4(b). This frame holds the base panel. A fabric layer attached to it acts as a growing media and no soil substrate is required. The plants are rooted in this fabric layer, and hydroponic technology is used to deliver water and nutrients to the plants. Usually, this is accompanied by waterproof membrane layers that protect the building from moisture (Medl et al., 2017; Palermo & Turco, 2020).

Another type of green wall, a "linear" green wall, has been suggested by some studies (Medl et al., 2017; Mitterboeck & Korjenic, 2017). Some studies even consider it as another class of VGS, i.e., pot-based (Hoffmann et al., 2023). In linear green walls, cascading elements are arranged in a linear fashion as shown in Fig. 7.4(c). These include linear planter boxes or pots, which are filled with the growth substrate and are arranged one above the other (Medl et al., 2017; Palermo & Turco, 2020).

7.3 Benefits of VGS

Vertical greening systems provide many environmental, health, societal, aesthetic, and energy-saving benefits (Ghazalli et al., 2019; Hoffmann et al., 2023). These benefits are not just enjoyed by individuals residing inside the building but also by the general public in and around that building. These benefits can be broadly classified under two headings, "Environmental Benefits" and "Health and Societal Benefits," for ease of understanding.

7.3.1 Environmental Benefits

Vertical greening systems result in the passive thermal regulation of a building and its surroundings. This leads to a reduction in the overall energy demand required for heating or cooling in order to maintain the optimum ambient temperature for human comfort, thus saving energy. VGSs help mitigate the urban heat island (UHI) effect. UHI is a metropolitan area that is significantly warmer than its neighboring areas as a result of the re-emission and trapping of heat by the asphalt and concrete covers in cities. The layer of vegetation provided by VGSs helps counter the

UHI effect and leads to cooling by shading and evapotranspiration (Price et al., 2015).

VGSs also reduce the greenhouse gas effect and air pollution both indoors and outdoors (Ghazalli et al., 2019; Suárez-Cáceres et al., 2021). Plants help in reducing the excess carbon dioxide in the atmosphere by converting it into sugars and water. Usually, plants with more woody stems are preferred for higher carbon sequestration (Charoenkit & Yiemwattana, 2017; Marchi et al., 2015). As well, the leaves act as an effective filter to trap the particulate matter (PM) suspended in air (Ottelé, 2011; Perini et al., 2017). VGSs have the potential for better retention of rainwater, and can also be effectively utilized for efficient greywater and stormwater management (Jin et al., 2018; Pearlmutter et al., 2021; Pucher et al., 2022). They also help increase the biodiversity in cities and may act as a measure for countering habitat fragmentation (Madre et al., 2015). However, detailed studies are required in this regard (Medl et al., 2017).

7.3.2 Health and Societal Benefits

VGSs have a positive impact on the health of a city's inhabitants. They can help reduce heat stress. Studies have shown that VGSs aid in reducing the universal thermal climate index (UTCI) and predicated mean vote (PMV) indicator values to optimum levels and, thus, increase human comfort and well-being (Hong et al., 2019; Medl et al., 2017). Interestingly, they can also help protect our eyes. In urban areas, there is increased glare due to the reflection and scattering of sunlight from the concrete/asphalt covers such as buildings, roads, pavement, etc. This can be potentially harmful to the eyes, especially due to the reflection and scattering of UV rays. Vegetation on walls and facades controls the intensity of reflected sunlight, thus protecting our eyes from glare and UV rays (Medl et al., 2017). VGSs may also aid people in coping with psychological stress and assist in faster recovery since they imitate nature (Hong et al., 2019). Furthermore, VGSs help in improving noise attenuation and buffering due to the absorption, reflection, and diffraction of sound by the vegetation layer (Bakker et al., 2023). They can also help protect buildings from climatic stresses, decrease their maintenance costs, and increase the value of the properties (Hoffmann et al., 2023). Finally, VGSs can contribute to the enhancement of the beauty of their neighborhoods.

7.4 Downsides/Challenges

Just like any other technology, VGSs have their disadvantages and limitations. VGSs are still an evolving technology and the cost of construction is pretty high, especially for wall-based systems with newer innovations and improvements. In addition, green facades and walls involve living elements that need regular maintenance for proper functioning. Lack of sufficient maintenance can lead to deterioration of the system, thus rendering it inefficient and unsustainable, both in terms of the environment and cost (Hoffmann et al., 2023; Huang et al., 2019). A cost-benefit analysis (CBA) performed by Perini and Rosasco (Perini & Rosasco, 2013) found only direct green facades to be economically sustainable due to their low installation, maintenance, and disposal costs. A detailed life-cycle analysis (LCA) by Ottele. et al (Ottelé et al., 2011) also show that none of the VGSs, except direct green facades, are sustainable, since the environmental and financial burden is more than the energy savings and other benefits. This burden can be reduced by suitable sustainable material selection and better envelope design. However, as technology continues to evolve and with the involvement of local communities, the cost of construction and maintenance of VGSs should come down with time. Apart from this, there are concerns regarding fire safety and controlling the insects and other small animals that may inhabit the systems and can cause problems for the inhabitants of the buildings (Hoffmann et al., 2021; Madre et al., 2015).

7.5 Thermal Regulation due to VGS

As mentioned before, VGSs have a significant impact on the energy transfer of buildings. In hot weather, they lead to a cooling effect, and, in cold weather, they result in warm interiors. The cooling effect is attributed to the shading effect of the canopy and the evapotranspiration caused by vegetation. In addition, the reflection and absorption of solar radiation by the green layer, and the insulating effect of the water and substrate (in the case of wall-based systems), contribute to this cooling effect. The green layer also prevents warm ambient air from penetrating inside the building (Bakhshoodeh et al., 2022; Šuklje et al., 2016). The warming effect of VGSs is mainly due to the re-emission of heat by the vegetation and the insulating effect of the green layer and the air trapped between the wall and the green layer (Convertino et al., 2022). Thus, VGSs can lead to building energy savings in all seasons.

7.5.1 Factors Affecting Thermal Regulation

7.5.1.1 Ambient weather conditions

Ambient temperature and weather conditions play a significant role in governing the thermal performance of a VGS. The same VGS can behave very differently depending on the weather it experiences. The greatest thermal impact is observed under the most extreme weather, i.e., the greatest cooling effect can be observed on the hottest day while the maximum heating effect is encountered on the coolest night. The amount of solar radiation, sky conditions (whether it is cloudy, rainy, sunny, etc.), humidity, water vapor pressure in the soil, wind, and airflow all affect the performance of a VGS significantly, since they directly influence the evapotranspiration and radiative and convective heat transfer rates.

7.5.1.2 Type and construction of VGS

Thermal performance varies with the type of VGS and its construction due to different configurations of various layers and the materials used. This difference is notable, especially when we compare green facades and green walls, since green walls also have a layer of substrate and water that can result in better cooling. Also, direct facades do not have an air gap between the wall and vegetation layer, unlike the other classes of VGS. The various layers present in a VGS, and their width and construction materials, all affect the thermal resistance of the VGS and, thus, the heat transfer rates.

7.5.1.3 Leaf area index

In the case of VGSs, the leaf area index (LAI) is defined as the ratio of the one-sided leaf area to the total vertical wall area. A pictorial representation of LAI is depicted in Fig. 7.5.

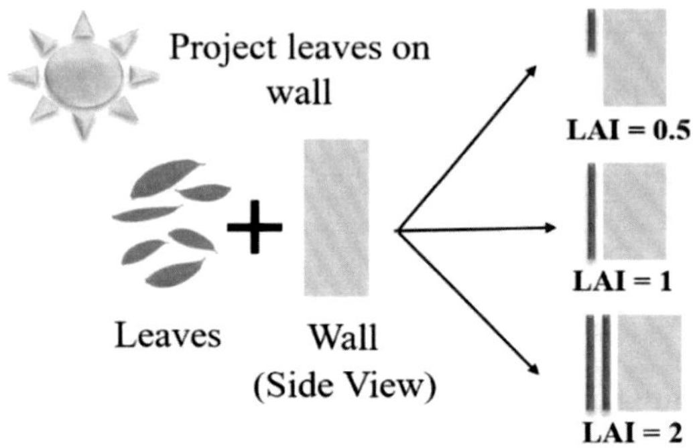

Figure 7.5 Pictorial depiction of leaf area index.

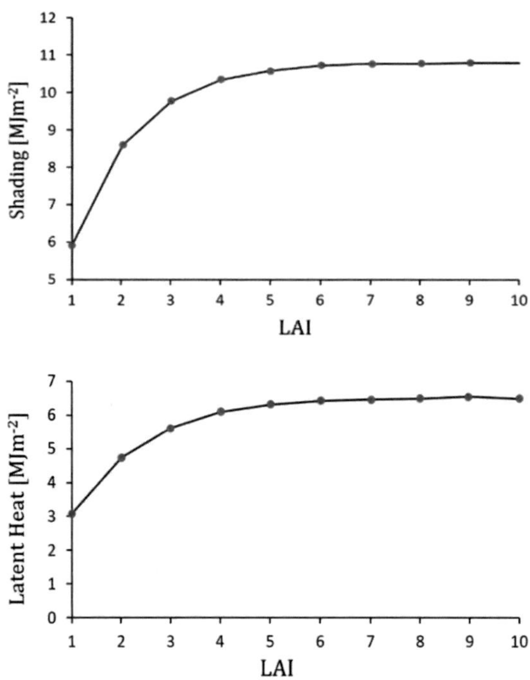

Figure 7.6 Variation of latent heat and shading with LAI. Adapted from a study by Convertino et. al at the University of Bari in 2022. (Convertino et al., 2022).

LAI can be thought of as a measure of coverage of the vertical wall by the VGS. It is a very important factor in determining the thermal performance of the VGS, because it directly affects the evapotranspiration rate and shading due to the coverage of the leaves. Usually, it has been observed that the cooling effect increases as the LAI increases. Figure 7.6 shows the trends of variation of latent heat due to evapotranspiration and shading effect with the LAI. A study conducted by Convertino et al. at the University of Bari (Convertino et al., 2022) shows that evapotranspiration rates and shading increase with an increase in LAI, up to a value of 6, and then almost become constant.

7.5.1.4 Air gap width

Most vertical greening systems have an air gap between the wall and the vegetation layer. In the case of direct VGSs, the width of this air gap can be assumed to be zero. A strong relationship has been observed between the gap width and the cooling effect. Usually, as the gap increases the cooling performance increases (Bakhshoodeh et al., 2022). However, the range of gap widths used in studies is limited, and further research is required to optimize the gap width for maximum thermal performance for a given VGS.

7.5.1.5 Facade orientation

The orientation of the facade or wall on which a VGS is mounted also contributes to the thermal performance, as it determines the amount and duration of solar exposure and shading. This is important for regulating plant growth as well. According to a review done by Bakhshoodeh et al. (2022), the cooling effect in the northern hemisphere during the summer is maximum for south-facing facades, followed by west-, east-, and north-facing facades, respectively. The order is reversed for the southern hemisphere.

7.5.1.6 Plant and substrate characteristics

Plant and substrate characteristics cannot be ignored when it comes to the thermal performance of VGSs. Transpiration and latent heat flux are dependent on the plant morphology. For example, studies have shown that the evapotranspiration rates and cooling performance for grasses are 3–4 times higher than for succulents (Bakhshoodeh et al., 2022). Cooling due to evapotranspiration also depends directly on the volumetric water content of the substrate. However, further studies are required to investigate and compare the effect of plant type and substrate characteristics on the performance of VGSs.

7.6 Mathematical Modeling

Mathematical modeling of VGSs is a very challenging problem because the systems involve many variables and living elements that interact with their surroundings, which can be complex to model. Since controlled experiments are difficult to perform, mathematical modeling and simulations become important to predict building energy requirements, as well as to optimize the thermal performance of the building and maximize energy savings.

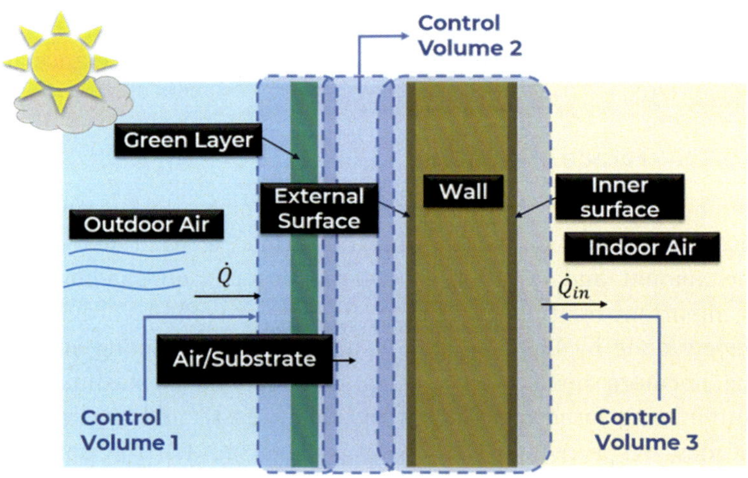

Figure 7.7 Simplified general representation of a VGS with the different regions and control volumes.

In order to simplify and understand the problem, while taking into account the practical construction of a VGS in general, the system has been divided into seven regions, as shown in Fig. 7.7.

These regions include the outdoor air; green layer; air gap (in the case of green facades) or a substrate layer (for green walls) depending upon the type of VGS; the external surface of the wall; wall; inner surface of the wall; and indoor air. Further, three control volumes have been considered for analysis, as shown in Fig. 7.7.

7.6.1 Control Volume 1

The first control volume includes the outdoor air, green layer, and air/substrate layer. Due to the involvement of the green layer, there are multiple interactions with the environment, which makes the analysis complex. A detailed description of all the interactions is shown in Fig. 7.8.

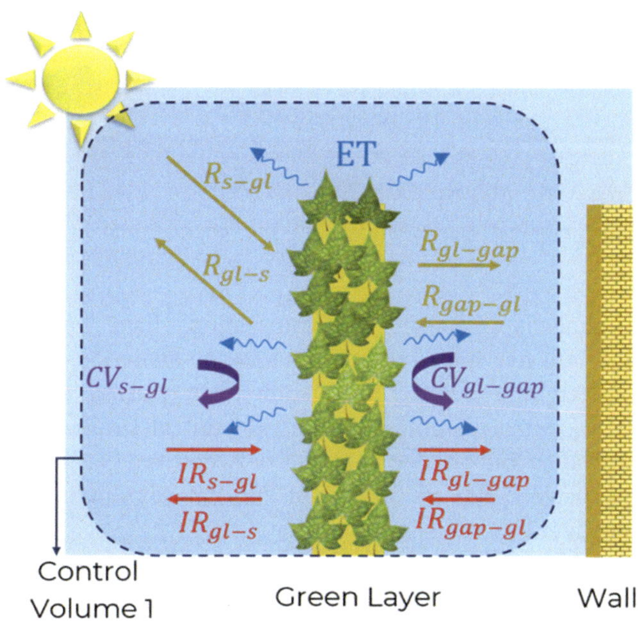

Figure 7.8 Interactions of the green layer with the environment. *R*, *IR*, *CV*, and *ET* refer to shortwave radiation, longwave radiation, convective heat transfer, and evapotranspiration energy exchanges, respectively. The subscripts '*s*', '*gl*', and '*gap*' denote the ambient surroundings, green layer, and air/substrate gap, respectively, and indicate the direction of energy exchange. For example, CV_{s-gl} refers to convection between surroundings and the green layer.

These include reflection, absorption, and re-emission of solar and longwave radiation by the vegetation layer, convection between the ambient outdoor air and green layer as well as between the green layer and air gap/substrate layer, and evapotranspiration from the plants.

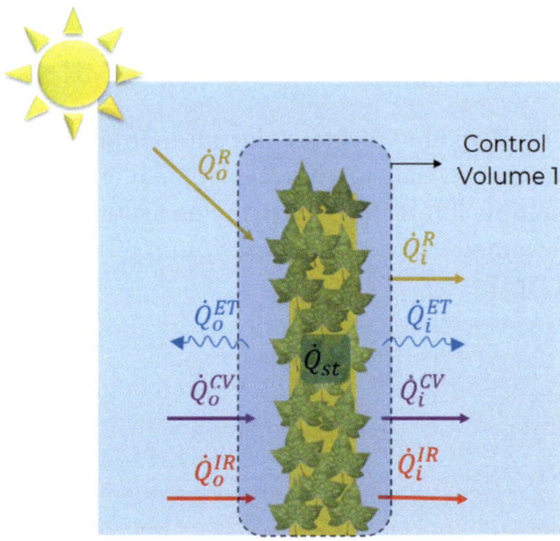

Figure 7.9 Simplified representation of the energy exchange in control volume 1.

For analysis purposes, the interactions have been further simplified into net solar radiation, longwave radiation, convection heat transfer, and evapotranspiration heat transfer entering or exiting the control volume through the control surfaces between the green layer and outdoor air and between the green layer and the air/substrate gap, as shown in Fig. 7.9. Conduction through the green layer has been neglected for simplicity.

Here, $\dot{Q}_o^R$, $\dot{Q}_o^{IR}$, $\dot{Q}_o^{CV}$, and $\dot{Q}_o^{ET}$ represent the rate of energy transfer across the control surface between outdoor air and the green layer, due to shortwave radiation, longwave radiation, convection, and evapotranspiration, respectively. Similarly, $\dot{Q}_o^{IR}$, $\dot{Q}_o^{IR}$, $\dot{Q}_o^{CV}$, and $\dot{Q}_o^{ET}$ depict the energy transfer rates across the control surface between the green layer and the air/substrate gap, due to shortwave radiation, longwave radiation, convection, and evapotranspiration, respectively. Some amount of energy is also stored by the plants, the rate of which is indicated by $\dot{Q}_{st}$.

Thus, from energy conservation in the control volume, we can write:

$$\dot{Q}_o^R + \dot{Q}_o^{IR} + \dot{Q}_o^{CV} - \dot{Q}_o^{ET} - (\dot{Q}_i^R + \dot{Q}_i^{IR} + \dot{Q}_i^{CV} + \dot{Q}_i^{ET}) = \dot{Q}_{st}. \tag{7.1}$$

For the steady-state operation of the system, $\dot{Q}_{st}$ can be assumed to be zero. Furthermore, there would be differences depending on whether green facades or walls are being considered. In the case of a green facade, there is an air gap between the green layer and the wall; thus, we can consider $\dot{Q}_o^{ET} = \dot{Q}_i^{ET}$ for simplicity, whereas for green walls, there is a layer of substrate between the green layer and the wall, instead of an air gap. Thus, all the heat flux due to evapotranspiration can be assumed to be in the direction of outdoor air from the green wall and, hence $\dot{Q}_i^{ET} \approx 0$.

7.6.2 Control Volume 2

The second control volume is located within the air gap or substrate and serves as a connecting link between control volumes 1 and 3, which can be seen in Fig. 7.7. The energy exchange in this control volume is shown in Fig. 7.10. $\dot{Q}_g$ is the amount of heat entering the control volume from the green layer and $\dot{Q}_{in}$ is the energy exiting the control volume into the indoor air. Some amount of energy is lost from the control volume, due to natural convection and radiation in the form of $\dot{Q}_{loss}$.

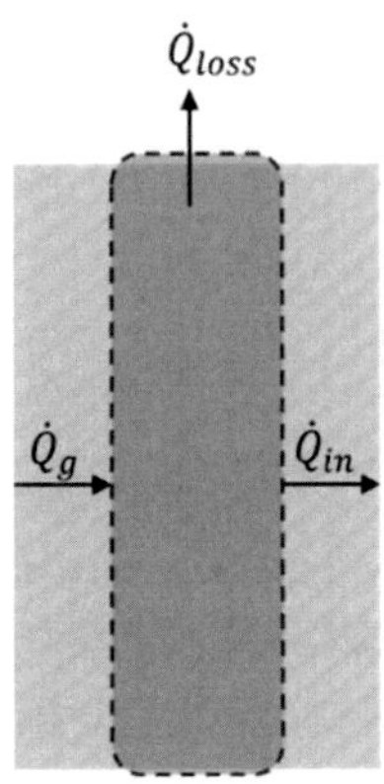

Figure 7.10 Energy exchange in control volume 2.

Considering that no amount of heat is stored in the air or substrate layer in a steady state, we can write

$$\dot{Q}_{in} = \dot{Q}_g - \dot{Q}_{loss}. \tag{7.2}$$

7.6.3 Control Volume 3

Control volume 3 includes the air or the substrate layer, the wall and its external and inner surfaces, and the indoor air. There is convective heat transfer occurring between the air gap/substrate layer, followed by conduction through the layers of the wall, and convection between the inner surface of the wall and the indoor air. Radiation from the wall is significantly smaller; hence, it can be neglected for simplicity. A detailed, simplified description is shown in Fig. 7.11.

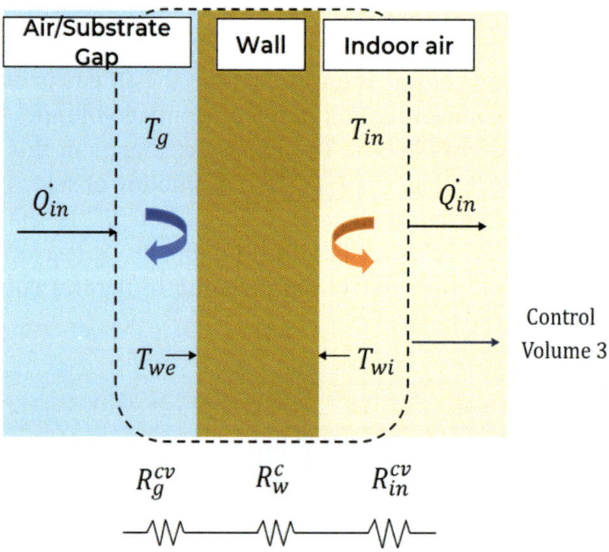

Figure 7.11 Heat transfer through control volume 3.

One-dimensional heat transfer can be assumed to be true in this control volume, since the area of the wall is generally much larger than its thickness. $\dot{Q}_{in}$ is the heat transfer through the control volume. T_g, T_{we}, T_{wi}, and T_{in} are the temperatures of the air/substrate gap, external wall surface, inner wall surface, and indoor air, respectively. Therefore, we can write

$$\dot{Q}_{in} = \frac{(T_g - T_{in})}{R_{tot}}, \tag{7.3}$$

where R_{tot} is the net thermal resistance of the system. Here, R_g^{cv} and R_{in}^{cv} are the thermal resistances, due to convection between the wall and air/substrate gap, and the wall and the indoor air, respectively. R_w^c is the thermal resistance of the wall in conduction. Since the thermal resistances are in series, it can be written as follows:

$$R_{tot} = R_g^{cv} + R_w^c + R_{in}^{cv}. \tag{7.4}$$

The thermal resistance of a wall depends on its composition. The simplest case is a composite wall made up of three layers—two layers of plaster and a layer of brick sandwiched in between, which is depicted in Fig. 7.12.

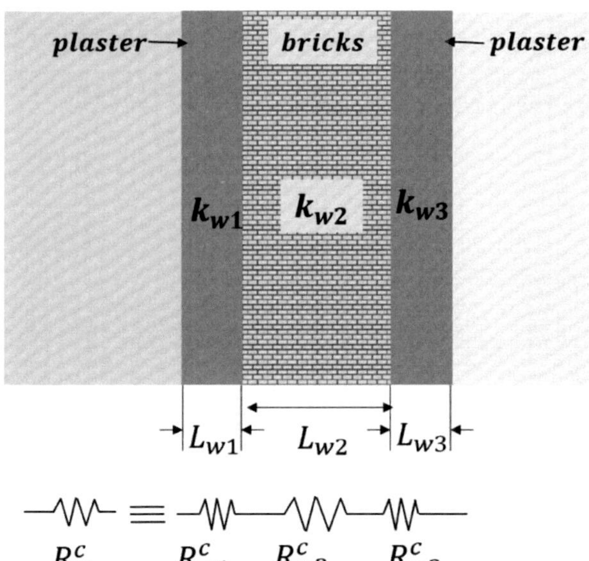

Figure 7.12 Equivalent thermal resistance of a composite wall.

Thus, the equivalent thermal resistance of this composite wall can be given as:

$$R_w^c = R_{w1}^c + R_{w2}^c + R_{w3}^c, \tag{7.5}$$

where R_{w1}^c, R_{w2}^c, and R_{w3}^c are the thermal resistances of the composite wall layers in conduction. If A is the cross-sectional area of the wall; L_{w1}, L_{w2}, and L_{w3} are the thicknesses; and k_{w1}, k_{w2}, and k_{w3} are the thermal conductivities of the layers, the thermal resistances of the composite wall layers in conduction become

$$R_{w1}^c = \frac{L_{w1}}{k_{w1}A}, \tag{7.6}$$

$$R_{w2}^c = \frac{L_{w2}}{k_{w2}A}, \tag{7.7}$$

and

$$R_{w3}^c = \frac{L_{w3}}{k_{w3}A}. \tag{7.8}$$

The thermal resistances in convection are given as

$$R_g^{cv} = \frac{1}{h_g A} \tag{7.9}$$

and

$$R_{in}^{cv} = \frac{1}{h_{in} A}. \tag{7.10}$$

However, the challenge here is to determine the values of the convection coefficients, h_g and h_{in}. The net energy transferred to the indoor air from outside can be calculated as:

$$\dot{Q}_{in} = \frac{(T_g - T_{in})A}{\left(\frac{1}{h_g} + \frac{L_{w1}}{k_{w1}} + \frac{L_{w2}}{k_{w2}} + \frac{L_{w3}}{k_{w3}} + \frac{1}{h_{in}}\right)}. \tag{7.11}$$

The major challenge in accurately modeling the interactions in a VGS is the dynamic and unpredictable nature of the system as well as the lack of properly controlled studies. Thus, while transitioning from the simplified mathematical model to real-world applications, the practical engineering limitations, the unpredictability of the system, and the dynamic interactions of the green layer with its environment must be carefully considered.

7.7 Discussion

Several studies have been conducted on vertical greening systems. However, there is much variation in terms of study parameters and conditions such as geographic location, climate zones, weather, system elements, collected data, etc., across different studies. Often, some of the parameters that can affect the thermal performance of a VGS are not reported in the studies. In such a situation, it is very difficult to compare the results of different studies except to get a general trend, since the system is dynamic, and many variables are involved. Additionally, there are simultaneous variations of multiple parameters while conducting experiments/studies due to the stochastic and dynamic nature of some of the critical parameters involved, such as weather conditions, leaf coverage, plant health, etc. This makes it more difficult to identify the exact relationship between thermal regulation and energy savings and the factors affecting them. These complications, coupled with the fact that there is no standard method for quantifying the performance of VGSs, make the comparison of different VGSs subjected to varying conditions next to impossible.

If VGSs are to be implemented on a large scale as a future solution to reduce the usage, or even replace, air conditioners and heating systems, it becomes very important that different VGSs can be compared, and decisions can be made as to which one works the best in a given location and under specific conditions. Thus, standardized universal performance indicators that can be used to compare VGSs in different locations, with different components, architectures, and costs are needed. Keeping this in mind, the following indicators are suggested, as shown in Fig. 7.13.

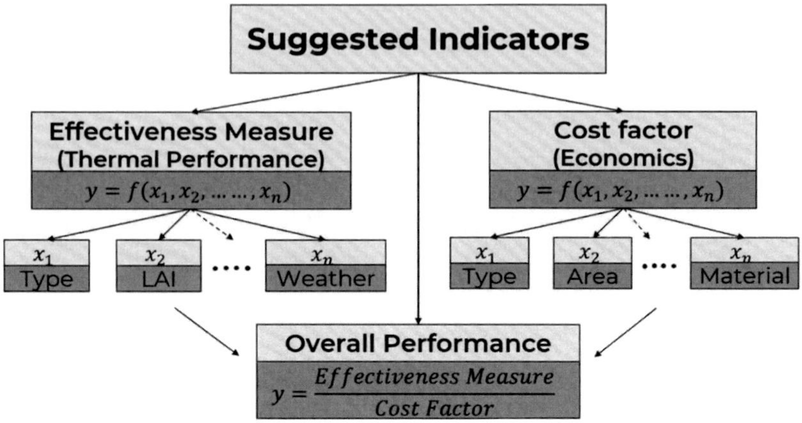

Figure 7.13 Suggested indicators for comparison of VGSs.

1. **Effectiveness Measure:** The effectiveness measure can be considered a measure of the thermal performance of the per unit area of the VGS. It depends on several independent variables—the type and construction of VGS being considered, its orientation, geographic location, LAI, plant species, and substrate considered. However, it should be noted that this indicator should be independent of any specific weather conditions, even though the performance of a VGS largely depends on the ambient weather. Since this is a generalized indicator that compares two or more different VGSs, it should be applicable to all weather conditions for a particular location. There can also be an effectiveness measure particularly for heating load reduction or cooling load reduction, to be more specific to the use case.
2. **Cost Factor:** The cost factor is a measure of the economic feasibility of the VGS. It is a factor that takes into account the cost per unit area of building and maintaining the VGS. It is influenced by the availability of resources, construction, plant species used, etc.
3. **Overall Performance Factor:** Usually, consumers take into account the performance and the cost of the system while comparing different VGSs. Hence, an overall performance factor should be considered, which takes into account both

the "effectiveness measure" and the "cost factor," to give a clearer picture. It can be defined as:

$$\text{Overall performance} = \frac{\text{Effectiveness measure}}{\text{Cost factor}}.$$

Obtaining a mathematical expression for these universal performance indicators is a very challenging task. Controlled studies that vary one parameter at a time are required to identify a functional relationship between the suggested universal performance indicators and the factors on which they depend. For this, it is important to devise methods to standardize the green infrastructure as standardization becomes crucial to obtain the values of the corresponding parameters and constants involved in the expression. Controlled studies can be performed for different types of VGSs with a specific type of vegetation in various locations and the results can be compiled to get a reference chart or a data book, which would be beneficial for designing VGSs in the future. However, a major limitation is that, often, performing controlled experiments can become very difficult and cumbersome considering the number of variables involved and the dynamic nature of the system. Mathematical studies, numerical simulations, and data-driven studies can be helpful in this regard.

It is to be noted that factors such as plant life cycle, health, population dynamics, and their interaction with the environment and other living organisms including insects are often ignored in studies for the sake of simplification. However, these factors can become important when it comes to the healthy functioning of the entire system over time. Since the system involves living elements and continuous interactions with the dynamic environment, it is very difficult to predict the exact behavior of the VGS. Additionally, opportunities such as floriculture or food production may be identified within the system.

7.8 Concluding Remarks

Vertical greening systems are an innovative future solution for reducing the negative effects of deforestation and urbanization, providing several other benefits at the same time. They help in passive temperature regulation and increase human comfort resulting in building energy savings in all seasons. VGSs are complex and dynamic systems that are challenging to model since they consist of living elements. Nevertheless, it is important to understand the systems in order to optimize their design for maximum thermal performance and energy savings. Controlled studies, varying one parameter at a time, are required to realize the functional relationship of the thermal performance with the design parameters. Further, there is a necessity to devise a methodology to standardize the system in order to define the universal performance indicators for comparing performances of different VGSs. Additionally, an attempt has been made to define the indicators; however, extensive studies are required to identify the mathematical expressions for the same.

References

Bakhshoodeh, R., Ocampo, C., & Oldham, C. (2022). Thermal performance of green façades: Review and analysis of published data. *Renewable and Sustainable Energy Reviews, 155,* 111744. https://doi.org/10.1016/j.rser.2021.111744.

Bakker, J., Lugten, M., & Tenpierik, M. (2023). Applying vertical greening systems to reduce traffic noise in outdoor environments: Overview of key design parameters and research methods. In *Building Acoustics.* https://doi.org/10.1177/1351010X231171028.

Charoenkit, S., & Yiemwattana, S. (2017). Role of specific plant characteristics on thermal and carbon sequestration properties of living walls in tropical climate. *Building and Environment, 115,* 67–79. https://doi.org/10.1016/j.buildenv.2017.01.017.

Convertino, F., Blanco, I., Vox, G., & Schettini, E. (2022). Thermal behaviour of green façades in winter climatic conditions. *International Journal of Environmental Impacts, 5*(1), 38–50. https://doi.org/10.2495/EI-V5-N1-38-50.

Convertino, F., Schettini, E., Blanco, I., Bibbiani, C., & Vox, G. (2022). Effect of leaf area index on green facade thermal performance in buildings. *Sustainability (Switzerland), 14*(5), 2966. https://doi.org/10.3390/su14052966.

Ghazalli, A. J., Brack, C., Bai, X., & Said, I. (2019). Physical and non-physical benefits of vertical greenery systems: A review. *Journal of Urban Technology, 26*(4), 53–78. https://doi.org/10.1080/10630732.2019.1637694.

Hoffmann, K. A., Schröder, S., Nehls, T., Pitha, U., Pucher, B., Zluwa, I., & Gantar, D. (2023). *Vertical Green 2.0 - The Good, the Bad and the Science.* Universitätsverlag der TU Berlin. https://doi.org/10.14279/depositonce-16619.

Hoffmann, K. A., Šuklje, T., Kozamernik, J., & Nehls, T. (2021). Modelling the cooling energy saving potential of facade greening in summer for a set of building typologies in mid-latitudes. https://doi.org/10.14279/depositonce-10512.

Huang, Z., Lu, Y., Wong, N. H., & Poh, C. H. (2019). The true cost of "greening" a building: Life cycle cost analysis of vertical greenery systems (VGS) in tropical climate. *Journal of Cleaner Production, 228*, 437–454. https://doi.org/10.1016/j.jclepro.2019.04.275.

Jin, Z., Xie, X., Zhou, J., Bei, K., Zhang, Y., Huang, X., Zhao, M., Kong, H., & Zheng, X. (2018). Blackwater treatment using vertical greening: Efficiency and microbial community structure. *Bioresource Technology, 249*, 175–181. https://doi.org/10.1016/j.biortech.2017.09.176.

Köhler, M. (2008). Green facades-a view back and some visions. *Urban Ecosystems, 11*(4), 423–436. https://doi.org/10.1007/s11252-008-0063-x.

Hong, K.-P., Lee, H., & Hong, S. (2019). A study on the physiological and psychological stress relief effects of vertical gardens on human body: 3 different construction methods of vertical gardens. *IEEE Xplore.* https://doi.org/10.1109/CISP-BMEI48845.2019.8966007.

Madre, F., Clergeau, P., Machon, N., & Vergnes, A. (2015). Building biodiversity: Vegetated façades as habitats for spider and beetle assemblages. *Global Ecology and Conservation, 3*, 222–233. https://doi.org/10.1016/j.gecco.2014.11.016.

Marchi, M., Pulselli, R. M., Marchettini, N., Pulselli, F. M., & Bastianoni, S. (2015). Carbon dioxide sequestration model of a vertical greenery system. *Ecological Modelling, 306*, 46–56. https://doi.org/10.1016/j.ecolmodel.2014.08.013.

Medl, A., Stangl, R., & Florineth, F. (2017). Vertical greening systems—A review on recent technologies and research advancement. In *Building and Environment*, 125, 227–239. https://doi.org/10.1016/j.buildenv.2017.08.054.

Mitterboeck, M., & Korjenic, A. (2017). Analysis for improving the passive cooling of building's surroundings through the creation of green spaces in the urban built-up area. *Energy and Buildings, 148,* 166–181. https://doi.org/10.1016/j.enbuild.2017.02.005.

Ottelé, M., Perini, K., Fraaij, A. L. A., Haas, E. M., & Raiteri, R. (2011). Comparative life cycle analysis for green façades and living wall systems. *Energy and Buildings, 43*(12), 3419–3429. https://doi.org/10.1016/j.enbuild.2011.09.010.

Ottelé, M. (2011). *The green building envelope: Vertical greening.* TU Delft, http://resolver.tudelft.nl/uuid:1e38e393-ca5c-45af-a4fe-31496195b88d.

Palermo, S. A., & Turco, M. (2020). Green Wall systems: Where do we stand? *IOP Conference Series: Earth and Environmental Science, 410*(1), 012013. https://doi.org/10.1088/1755-1315/410/1/012013.

Pearlmutter, D., Pucher, B., Calheiros, C. S. C., Hoffmann, K. A., Aicher, A., Pinho, P., Stracqualursi, A., Korolova, A., Pobric, A., Galvão, A., Tokuç, A., Bas, B., Theochari, D., Milosevic, D., Giancola, E., Bertino, G., Castellar, J. A. C., Flaszynska, J., Onur, M., ... Nehls, T. (2021). Closing water cycles in the built environment through nature-based solutions: The contribution of vertical greening systems and green roofs. *Water (Switzerland), 13*(16), 2165. https://doi.org/10.3390/w13162165.

Perini, K., Ottelé, M., Giulini, S., Magliocco, A., & Roccotiello, E. (2017). Quantification of fine dust deposition on different plant species in a vertical greening system. *Ecological Engineering, 100,* 268–276. https://doi.org/10.1016/j.ecoleng.2016.12.032.

Perini, K., & Rosasco, P. (2013). Cost-benefit analysis for green façades and living wall systems. *Building and Environment, 70,* 110–121. https://doi.org/10.1016/j.buildenv.2013.08.012.

Price, A., Jones, E. C., & Jefferson, F. (2015). Vertical greenery systems as a strategy in urban heat island mitigation. *Water, Air, and Soil Pollution, 226*(8), 247. https://doi.org/10.1007/s11270-015-2464-9.

Pucher, B., Zluwa, I., Spörl, P., Pitha, U., & Langergraber, G. (2022). Evaluation of the multifunctionality of a vertical greening system using different irrigation strategies on cooling, plant development and greywater use. *Science of the Total Environment, 849,* 157842. https://doi.org/10.1016/j.scitotenv.2022.157842.

Raji, B., Tenpierik, M. J., & Van Den Dobbelsteen, A. (2015). The impact of greening systems on building energy performance: A literature review. *Renewable and Sustainable Energy Reviews, 45*, 610–623. https://doi.org/10.1016/j.rser.2015.02.011.

Suárez-Cáceres, G. P., Fernández-Cañero, R., Fernández-Espinosa, A. J., Rossini-Oliva, S., Franco-Salas, A., & Pérez-Urrestarazu, L. (2021). Volatile organic compounds removal by means of a felt-based living wall to improve indoor air quality. *Atmospheric Pollution Research, 12*(3), 224–229. https://doi.org/10.1016/j.apr.2020.11.009.

Šuklje, T., Medved, S., & Arkar, C. (2016). On detailed thermal response modeling of vertical greenery systems as cooling measure for buildings and cities in summer conditions. *Energy, 115*, 1055–1068. https://doi.org/10.1016/j.energy.2016.08.095.

Chapter 8

Evolutionary Structurally-Sound-Design-Process of Dovetail Massive Wooden Board Elements as a Sustainable and Environmentally Friendly Solution

Hüseyin Emre Ilgın,[a] Markku Karjalainen,[a] and Sami Pajunen[b]

[a]*School of Architecture, Faculty of Built Environment, Tampere University, Finland*
[b]*Civil Engineering, Faculty of Built Environment, Tampere University, Finland*

hemreilgin@gmail.com

In engineered wood products (EWPs), the inclusion of adhesives and metal fasteners is essential. However, adhesives, being petroleum-based, can result in the emission of harmful gases, and the presence of metal fasteners can hinder the reusability of these products. These considerations give rise to significant inquiries concerning the sustainability and ecological repercussions of EWPs. Consequently, there is a need for an alternative solution that relies solely on solid and pure wood, eliminating the need for adhesives and metal connectors. This solution is provided by dovetail massive wood board elements (DMWBEs). Despite an abundance of existing studies in the literature exploring the technical, environmental, and financial aspects of DMWBEs, there

Reclaiming Eden: Responsible Living, Engineering, and Architectures
Edited by David S.-K. Ting and Jacqueline A. Stagner

ISBN 978-981-5129-31-1 (Hardcover), 978-1-003-49612-0 (eBook)
www.jennystanford.com

is a dearth of research specifically assessing their technical performance. The primary objective of this chapter is to fill this research gap by presenting a range of structurally sound design solutions for lateral load-bearing elements in the construction of multistory buildings, employing architectural modeling programs. The suggested arrangements are grounded in a theoretical methodology, considering modern construction methodologies instead of relying solely on structural or mechanical models. Following this design phase, further research will be conducted, including technical performance tests, to evaluate the capabilities of DMWBEs. It is anticipated that this chapter will contribute to the wider promotion, adoption, and application of DMWBEs in pioneering architectural and structural contexts, specifically in the construction of multistory timber buildings.

8.1 Introduction

Wood, apart from being an environmentally conscious and sustainable construction material, possesses advantageous features such as a favorable strength-to-weight ratio, ease of manipulation, efficient thermal insulation, and commendable fire performance in substantial elements. Furthermore, wood offers recyclability and reusability benefits. Nonetheless, due to its innate properties, wood encounters certain limitations when employed in construction scenarios. The anisotropic characteristics of timber result in diminished mechanical characteristics in the transverse direction when contrasted with the longitudinal direction. Furthermore, in the same species and even within an individual tree, discrepancies in growth circumstances lead to a range of features such as grain angle and latewood proportion. The dimensional alterations caused by moisture further pose challenges in meeting the demands of contemporary timber construction [1, 2].

To overcome these obstacles, a variety of EWPs have been created and widely employed, such as CLT, glued-laminated timber (GLT), laminated veneer lumber (LVL), nail-laminated timber (NLT), and plywood. Typically, EWPs are fabricated through the adhesive bonding of solid wood, veneer, or wood fragments like particles, and fibers, or through the mechanical connection of timber blocks to form large structural elements. In contrast to traditional sawn timber, EWPs provide expanded opportunities in the building industry, demonstrate decreased alterability in characteristics, and offer improved dimensional stability [3]. Additionally, the utilization of EWPs in off-site prefabrication enhances the overall quality of the final building while also reducing construction time and costs [4]. EWPs are gaining traction, particularly in high-rise timber construction [5–10], as exemplified by notable structures like Mjøstårnet, standing at a height of 85 meters and consisting of 18 stories, in Norway, constructed in 2019 [11, 12] (refer to Fig. 8.1), HoHo, standing at a height of 84 meters and consisting of 24 stories, in Austria, completed in 2020 [13] (refer to Fig. 8.2), and Treet, standing at a height of 49 meters and consisting of 14 stories, in Norway, erected in 2015 [14].

Figure 8.1 Mjøstårnet (Norway, 2019) (Courtesy of DERFRITZ).

Figure 8.2 HoHo (Austria, 2020) (Courtesy of DERFRITZ).

Nonetheless, EWPs are primarily assembled using artificial glues or metal connectors, which have been proven to have high energy requirements during fabrication and, consequently, a substantial environmental impact. The embodied energy of GLT is about 60% higher than that of conventional sawn timber [15]. In their review of CLT in buildings, Cadorell and Crawford [16] recognized the benefits of CLT in reducing non-renewable material depletion, but they indicated that the scope of other ecological advantages remains uncertain. Additionally, the inclusion of adhesives and metal elements in EWP presents difficulties in terms of reusability and recyclability [17]. Puettmann et al. [18] stated that despite the adhesives constituting a mere 1.4% (by mass) of CLT, the entire life cycle of the adhesive contributes to approximately 31% of the environmental impact concerning global warming potential, total energy, and non-renewable resources, and approximately 57% for the eutrophication potential of the final CLT product. The expenses associated with adhesive bonding of lamellae in CLT form a substantial component of the total production cost, not solely owing to the glue material, but also due to equipment preservation and the time-intensive curing process unless expensive adhesive materials or specialized machinery such as high-frequency/microwave systems are employed. Moreover, the existence of bonding agents and metal fasteners may present difficulties in wood processing procedures such as cutting and planning. Throughout utilization, the adhesives found in EWPs have the potential to release volatile organic compounds, which may have consequences for both the environment and human health [19]. Emissions tend to increase particularly at elevated temperatures and high relative humidity levels [20]. Moreover, the adhesives and metal fasteners represent susceptible elements of EWPs in fire scenarios, as they have the potential to contribute to the early deterioration of wooden buildings [21–23].

Even though glue and metal fasteners are widely acknowledged for their dependability and extended operational capabilities, an increasing need exists for ecologically sustainable alternatives to connect wooden components in multi-layer timber beams and panels. Nowadays, diverse adhesive-free practices for joining such beams and panels have surfaced. These advancements necessitate a thorough evaluation of their individual strengths and weaknesses, including their technical limitations, manufacturing expenses, and mechanical characteristics.

Existing literature extensively covers the technical, environmental, and financial dimensions of EWPs within the building sector, encompassing various construction solutions. However, a notable research gap exists in evaluating the practical functioning of dovetail massive wooden board elements (DMWBEs) specifically in multi-story or high-rise structures [24]. Furthermore, the research on DMWBE is currently limited, with state-of-the-art studies primarily focusing on individual members or small-scale prototypes. The analysis remains constrained to connection details and adopts a predominantly theoretical approach, offering limited insights from a structural standpoint [25].

The primary focus of this study revolves around the exploration of ecologically sensitive engineering solutions through the implementation of dovetail massive wood board elements as sustainable alternatives (refer to Fig. 8.3). These elements, based on an ancient joining technique (refer to Fig. 8.4), offer a sustainable solution that is free from adhesives and metal fasteners. They consist of solid and entirely natural wood, promoting healthy indoor air quality. However, the exclusive use of dovetail elements, known as glueless boards, can present certain challenges and limitations, particularly regarding dimensional stability. The substantial decrease in equilibrium humidity content during the fabrication and assembly of the boards may result in elevated wood shrinkage and the creation of spaces between individual boards, potentially influencing wall stiffness in a negative manner. Additionally, specific configurations of DMWBE may result in reduced wood strength in the radial and tangential directions. Still, there is confidence that these potential concerns can be addressed by optimizing and refining the approach based on performance testing, encompassing evaluations of structural behavior, moisture transfer resistance, and airtightness.

Figure 8.3 DMWBE prototype.

Figure 8.4 The dovetail joint, one of the earliest joining methods used in woodworking.

Despite a plethora of existing literature delving into the technical, environmental, and financial facets of DMWBEs, there exists a scarcity of research that specifically evaluates their technical performance. The principal aim of this chapter is to address this research gap by introducing an array of structurally sound design solutions for lateral load-bearing components in the construction of multi-story structures, utilizing architectural modeling software. The proposed configurations are founded on a theoretical methodology that takes into account contemporary construction approaches, moving beyond sole reliance on structural or mechanical models. Subsequent to this design phase, further investigation will be undertaken, encompassing technical performance assessments, to appraise the capacities of DMWBEs. It is envisaged that this chapter will make a contribution to the broader propagation, adoption, and utilization of DMWBEs in pioneering architectural and structural contexts, specifically within the domain of multi-story timber building construction.

In this chapter, the designation "wood" or "timber" pertains to EWPs such as CLT. CLT is a prefabricated multi-layer EWP produced by bonding a minimum of three layers of boards together with adhesive under pressure. Additional instances of EWPs encompass glue laminated timber, which is produced by gluing graded timber laminations with their grain oriented parallel to the longitudinal axis of the section, and laminated veneer lumber, created by bonding thin vertical softwood veneers with their grain oriented parallel to the longitudinal axis of the section through the application of heat and pressure.

The organization of this chapter followed a systematic structure. Initially, an in-depth review of pertinent academic literature was conducted. Subsequently, the study's materials and methods were described, and the obtained results were presented. Ultimately, a thorough analysis and conclusive remarks were developed.

8.2 Historical Background of Dovetail Joint Technique

Throughout history, the dovetail method has been extensively utilized as a primary technique of woodwork, encompassing applications such as furniture making, cabinet and log construction, and traditional wooden structures. The historical roots of the dovetail joint method can be identified in antiquity, evident in ancient Egyptian furniture found in tombs dating back to the First Dynasty and in stone pillars discovered in Indian temples (refer to Fig. 8.5), and in the architectural traditions of Japanese and Korean structures [26, 27]. This technique also found its place in Chinese ancient architecture, where it was officially introduced as one of the fundamental joint methods during the Song Dynasty in 1103, as reflected in national building codes and construction practices [28–30]. Furthermore, throughout antiquity until the Middle Ages, Egyptian cabinet construction primarily utilized mortise and tenon, dovetail, and mitered joints [31]. In European woodworking tradition, the dovetail joint is recognized under different names, including alternatives like the swallowtail joint or fantail joint [32].

Residential constructions in the 13th century marked the beginning of wood-framed structures, featuring a range of joint types such as mortise and tenon joints, wedged joints, notched joints with tenons, and dovetail joints. Churches in the 14th century provide notable instances where the dovetail joint (refer to Fig. 8.6) was used to connect roof rafters and beams [33]. The roof framework of the St. Jacob's Church in Torun, which dates back to the 16th century, serves as one of the earliest existing instances of notched joints using dovetail tenons [34]. Additionally, the Cewków Church (see Fig. 8.7) and the Chotylub Church, both 19th-century Polish churches, stand as notable examples of timber-framed constructions showcasing dovetail wall-corner joists.

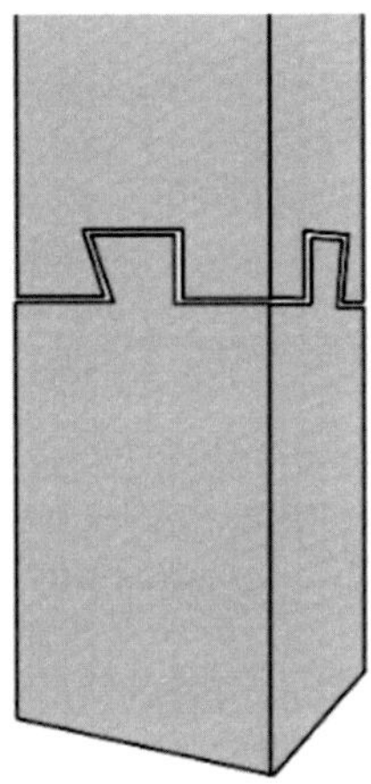

Figure 8.5 A temple in India featuring a stone pillar.

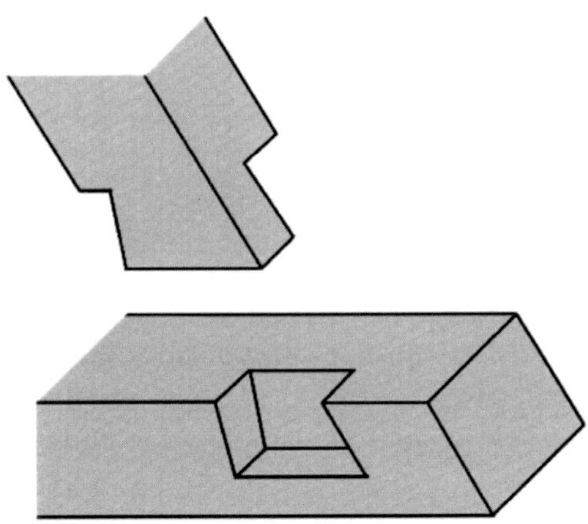

Figure 8.6 Dovetail joint.

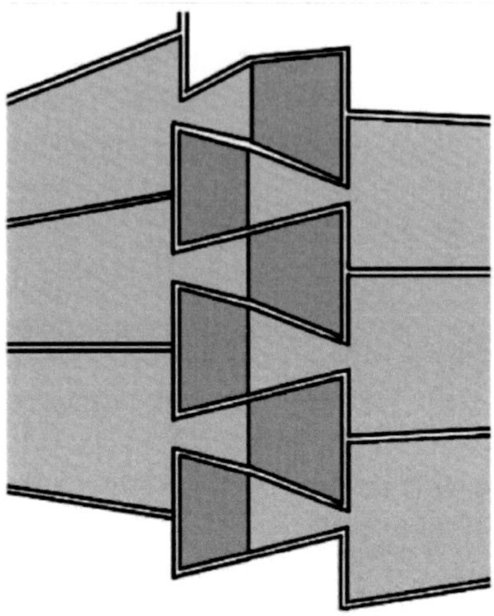

Figure 8.7 Dovetail joint detail of Church in Cewków, Poland.

Prior to the middle of the 20th century, wood-to-wood joints based on carpentry techniques were widely utilized in building construction owing to the proficiency of experienced woodworkers in design and production [35]. Practical considerations often dictated the different dovetail models in Europe and Asia [36]. Nevertheless, the high costs associated with labor and limitations arising from overly conventional designs rendered these joineries economically irrational. Nevertheless, progress in computer numerical control (CNC) wood processing machines has rejuvenated the cost-effectiveness of carpentry-type wood-to-wood joints.

8.3 Literature Review

To date, the existing body of literature has extensively covered the technical features of timber in the building industry, encompassing various solutions that incorporate EWPs like CLT [37–40]. Limited attention has been given to the exploration of novel adhesive- and metal-free assembly methods, such as wood dowels, and dovetail joining, as viable alternatives for constructing prefabricated EWPs [41–46]. Similarly, nevertheless, the body of research on dovetail massive wood elements (DMWEs) is relatively sparse, with only a few studies (e.g., [47]) dedicated to this topic. The current body of literature predominantly emphasizes specific aspects, such as conducting structural analyses and model testing on different joint configurations, without offering a holistic evaluation of the overall technical performance. As a result, there is a need for further investigation to evaluate the performance of DMWE in terms of structural integrity, fire resistance, sound transmission, and other relevant factors at both the component and system levels.

One of the significant studies in recent years pertaining to wood-to-wood connections, specifically dovetail wood joints, was conducted by Xie et al. [48]. Their investigation centered on analyzing the contact attributes of mortise and tenon joints in conventional timber constructions. Through the utilization of structural modeling software, they employed both simulations and experimental validations. The results indicated that the customized finite element simulation displayed a stronger correlation with the real force distribution and structural behavior of mortise and tenon joints.

In their study, Gamerro et al. [49] conducted a research study introducing an innovative idea for building components utilizing tenon joints. Their method involved the utilization of portable flat packs that could be transported and assembled on-site. The researcher's objective was to create a computational model capable of forecasting the semi-rigid behavior of the joints and the effective bending stiffness of these structural elements. The findings showed that the suggested simulation provided a realistic methodology for analyzing the stress distribution and overall displacements of interlocked components. However, additional research is required to ensure that the design of these components complies with building standards and aligns with the prevailing conditions in the building industry.

Sha et al. [50] carried out a study with the goal of evaluating the influence of damage to mortise-tenon joints on the cyclic behavior of a conventional Chinese wooden frame through finite element analysis. The research comprised subjecting the model to lateral cyclic loading and corroborating the outcomes with experimental tests. Three distinct forms of damage were identified and simulated, encompassing the separation between the mortise and tenon, as well as damage occurring at both the upper and lower extremities of the tenon. The findings demonstrated that these types of damage have detrimental impacts on the lateral behavior of the wooden frame, leading to reduced rigidity.

Jeong and Song [51] performed a comprehensive examination of the mechanical features of dovetail joints when subjected to tensile load, employing three different methodologies. The study focused on evaluating parameters such as initial stiffness and ductility ratio through these three methods. The findings highlighted that the inclination of the preliminary load-displacement curve is significantly influenced by the gap present in the dovetail joint. Moreover, yield displacement rates of the dovetail joints are substantially dependent on the original slope, which varies depending on the specific analytical model employed.

Branco and Descamps [52] introduced a range of carpentry joints, such as notched and scarf joints, along with calculation principles and potential strengthening procedures. The findings of their study can be summarized as follows:

(1) In situations where wooden components undergo substantial deterioration, replacement emerges as the sole feasible option.
(2) In scenarios necessitating repairs, dependable in situ assessment methods are utilized to ascertain the suitable extent of intervention.
(3) The inadequate research conducted in this domain is apparent, with the absence of precise findings and design principles for retrofitting old woodwork joints.
(4) To ensure expertise, engineers need specific utensils for designing dowelled joinery.

The study conducted by Jeong et al. [53] explored the influence of numerical factors on the structural performance of the dovetail joints and determined its maximum structural capability using the finite element model. Numerous stress distributions related to geometric variations were considered. The findings revealed that shear and perpendicular-to-grain tension stresses were the most significant factors affecting the connection's strength. Furthermore, the structural models were validated by comparing the estimated strength of the dovetail joint with the results obtained from experimental tests.

The objective of Ozkaya et al.'s study [54] was to investigate the impact of joint quantity and adhesive type on the diagonal tensile strength of frames made from the Oriented Strand Board. A total of 152 OSB samples were tested. The findings indicated that the use of glue in the corner joints of the dovetail connections is crucial for enhancing the diagonal tensile strength of the frame.

In addition to the aforementioned studies, other research (e.g., [55–58]) has primarily concentrated on conducting structural analyses and modeling analysis of diverse joint details in various geometric arrangements, rather than assessing the implementation of a load-bearing component, such as a shear wall.

8.4 Materials and Methods

The methodology for this chapter was meticulously crafted, employing a multi-faceted approach that integrates a comprehensive review of academic literature and sophisticated architectural modeling methodologies. This comprehensive methodology was structured to effectively address the research complexities and design intricacies inherent in architectural endeavors.

The foundational aspect of this research was built upon an extensive review of academic literature, focusing predominantly on internationally recognized peer-reviewed journals and analogous research initiatives. This literature review provided a strong theoretical underpinning and informed the research process, enabling a thorough understanding of existing knowledge and paving the way for novel insights and contributions.

To bridge the theoretical understanding with practical implementation, architectural modeling methodologies were crucially employed in the execution of this chapter. These modeling techniques served as indispensable tools, enabling architects to grapple with intricate research and design challenges effectively. By utilizing advanced architectural modeling methodologies, this research was able to conceptualize, document, discuss, and disseminate information in a tangible and coherent manner [54]. This facilitated a seamless transition from abstract concepts to practical design applications, aligning with the overarching objectives of this study.

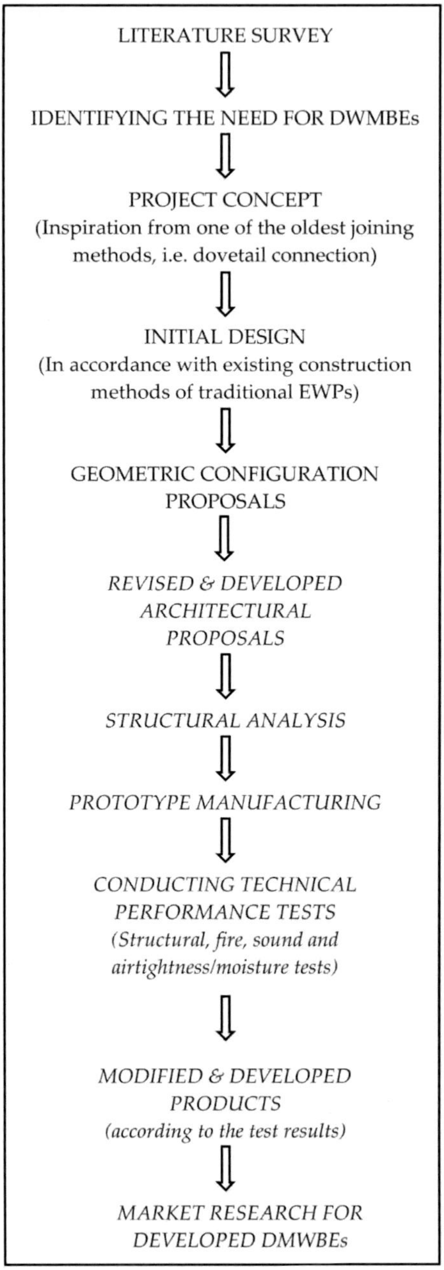

Figure 8.8 The steps of the initial design employed as a research approach in this investigation and the forthcoming stages of the project *(in italics)*.

Notably, architectural design research widely embraces and relies upon these modeling techniques as vital instruments in shaping the research approach [59, 60]. The absence of a singular standardized method for constructing the object and subject of architectural activity underscores the critical variations in research methodologies and architectural design strategies, especially during pivotal problem-solving phases [61]. Addressing this challenge, this research adeptly utilized prominent software applications such as AutoCAD and SketchUp, coupled with advanced object modeling methods commonly deployed in contemporary architectural design applications [62–65]. This choice of software and modeling methods was deliberate, aligning with current industry practices and optimizing the research process [66, 67].

In this setting, the research materialized novel suggestions through a synergistic combination of sketches and representations, presenting illustrative depictions that encouraged a dynamic and novel perspective on architectural research. Figure 8.8 was meticulously crafted to visually represent the steps of the initial design approach utilized in this research, followed by subsequent project phases. This visual representation serves as a guide, offering a clear and structured view of the research trajectory and design evolution, encapsulating the essence of this chapter.

8.5 Findings

The novelty of DMWBE is rooted in a fresh approach that combines a novel comprehension of wood characteristics and its potential, traditional woodworking expertise, efficient and precise large wood board milling abilities, and digital design. Consequently, architects, structural engineers, and production units can jointly engage in the same file, producing the intended result. The number of layers can be extensively adjusted, and the timber's width and thickness can also differ based on requirements, resulting in a fully formed board without the need for adhesives, nails, or any other size limitations, setting it apart from competing CLT and LVL products.

Referring to established construction practices of different EWPs, like CLT, as a preliminary step in design and implementation, innovative and architecturally sound 2D and 3D floor slab models are showcased below. In order to facilitate a meaningful comparison with CLT of similar sizes, the selected ideal test size for DMWBE will predominantly consist of the following dimensions: 200 mm thick (comprising 5 layers), 1450 mm wide, and 5075 mm long, serving as the longest test specimens for conducting structural bending tests. Nevertheless, it is crucial to recognize that the sizes of the structural elements might differ, especially when taking into account structural analysis, subsequent performance tests, and evaluations for fire safety, sound insulation, and airtightness under varying humidity conditions.

Prototypes of DMWBEs were created at Lapland Vocational College, located in Finland. The manufacturing of the two test specimens involved employing a 5-axis CNC machine (Fig. 8.9). To enable an integrated environment for various stages, including finishing, toolpath optimization, and G-code simulation for the manufacturing process, the CNC post-processor methodology was utilized. Throughout fabrication, the dovetail boards' moisture content ranged from 10% to 12%. However, challenges arose during the production process, primarily due to the lack of a mass production line, resulting in relatively extended production times, as well as the necessity for diverse tools like blades, along with the need to manage dust removal during manufacturing.

As illustrated in Fig. 8.10, the utilization of "solid/massive type" elements has emerged as a promising alternative to traditional slab flooring, demonstrating their potential to function as DMWBEs. This innovative concept finds its roots in the historical practice of employing dovetail, an age-old joining method showcased in Fig. 8.3, within revered architectural structures like temples and churches.

The inspiration for implementing the "solid/massive type" approach stems from an in-depth examination of ancient construction techniques, particularly the use of dovetail joints. Figure 8.3 exemplifies the intricate dovetail joinery present in temples and churches, showcasing the remarkable craftsmanship of past civilizations. This historical methodology involves precisely crafted interlocking wooden blocks, providing structural stability and durability.

Drawing from this historical knowledge, modern architecture embraces the concept of "solid/massive type" elements as a contemporary adaptation of traditional dovetail joinery. The integration of these elements into the construction of floors offers notable advantages, such as enhanced load-bearing capacity, increased structural resilience, and improved acoustic and thermal insulation.

Figure 8.9 The production of DMWBEs at Lapland Vocational College in Kemi, Finland, was carried out using a CNC machine with a 5-axis setup.

The "solid/massive type" approach involves crafting substantial wooden blocks or elements with dovetail-like interlocking features, akin to those observed in ancient constructions. These elements are meticulously designed to seamlessly fit together, ensuring a robust connection and distributing loads effectively. The precision and durability of this interlocking system significantly contribute to the overall stability and longevity of the flooring structure.

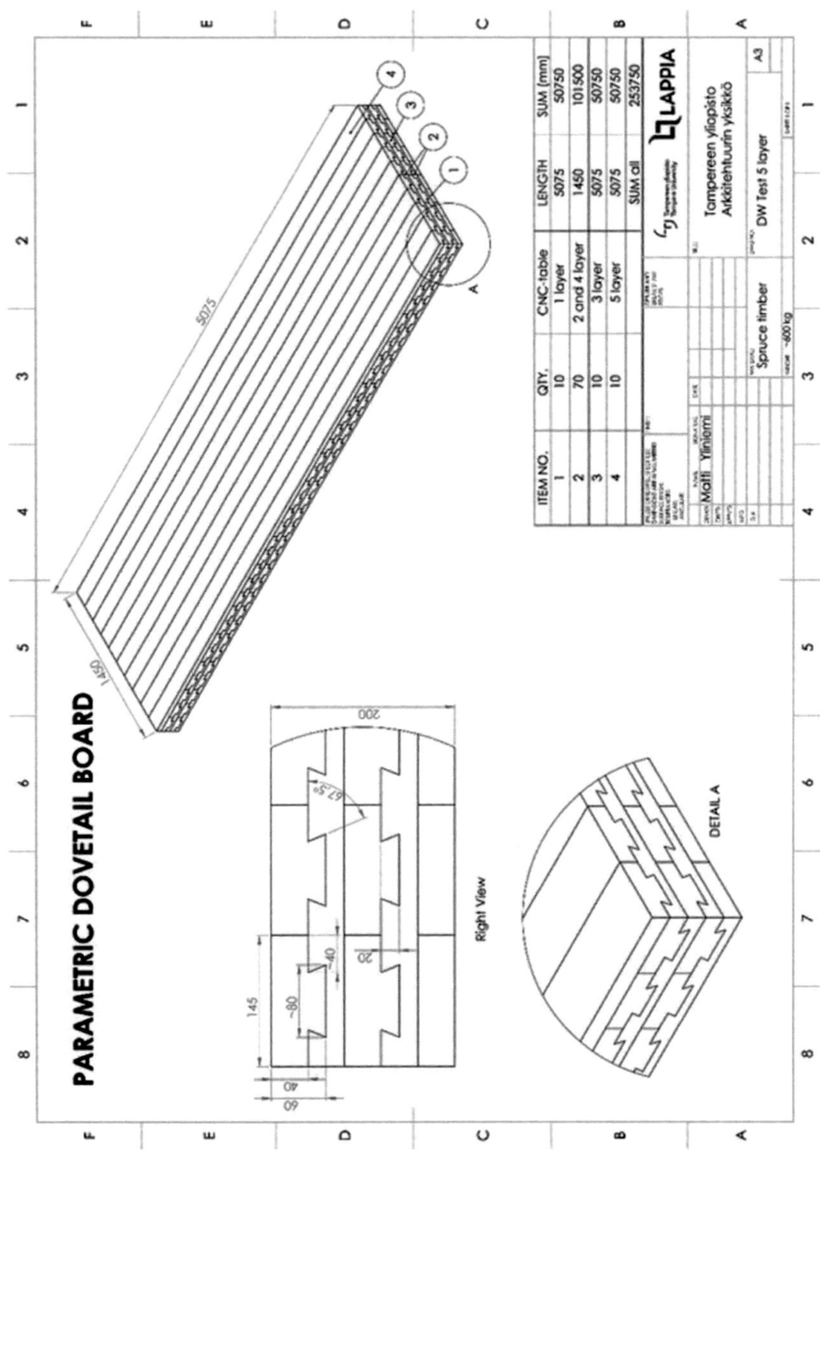
PARAMETRIC DOVETAIL BOARD
5075
1450
Right View
200
145
~80
40
20
60
67.5°
DETAIL A
ITEM NO. | QTY. | CNC-table | LENGTH | SUM (mm)
1 | 10 | 1 layer | 5075 | 50750
2 | 70 | 2 and 4 layer | 1450 | 101500
3 | 10 | 3 layer | 5075 | 50750
4 | 10 | 5 layer | 5075 | 50750
SUM all | 253750
LAPPIA
Tampereen yliopisto
Arkkitehtuurin yksikkö
Matti Yliniemi
Spruce timber
~600 kg
DW Test 5 layer
A3

Figure 8.10 DMWBEs as a substitute for floor slabs, specifically the solid/massive type.

Furthermore, the incorporation of DMWBEs, inspired by the dovetail concept, aligns with contemporary sustainable building practices. The use of wood, a renewable and eco-friendly material, underscores a commitment to environmentally responsible construction. This approach also encourages the utilization of locally sourced wood, promoting regional craftsmanship and reducing transportation-related carbon emissions.

As depicted in Fig. 8.11, the strategic integration of wooden dowels played a crucial role in augmenting the bending capacity of the board showcased in Fig. 8.10. This method involved inserting wooden dowels into the board, providing reinforcement and structural support. However, while this approach proved effective in enhancing bending resistance, it presented certain drawbacks that need to be considered.

The use of wooden dowels, while effective, introduced complexities into the manufacturing process. The incorporation of dowels necessitated meticulous precision and additional steps during production, contributing to an increase in manufacturing time and complexity. Crafting the holes for dowel placement and ensuring precise alignment demanded specialized skills and machinery, adding intricacy to the fabrication of the board.

Moreover, the implementation of wooden dowels required a higher degree of labor intensity. Skilled workers were essential to accurately position and secure the dowels within the board, adding to the overall labor costs and time investments. This heightened labor requirement might limit the scalability and cost-effectiveness of the approach, especially in large-scale manufacturing scenarios.

To mitigate these challenges and streamline the process, future research and development should focus on optimizing the manufacturing techniques associated with wooden dowel integration. Innovations in automated machinery and precision tools could potentially streamline dowel insertion, reducing labor intensity and enhancing production efficiency. Additionally, exploring alternative materials or methods that offer comparable reinforcement without the complexities linked to wooden dowels could be an avenue for further improvement.

Figure 8.11 Dovetail massive wood board elements reinforced with wooden dowels.

In the pursuit of fortifying the bending capacity of the panel presented in Fig. 8.10, an innovative alternative approach was devised, employing a triple-layer design. This design ingeniously harnesses the inherent strength of the middle layer, strategically positioning it to counteract bending forces, as vividly demonstrated in Fig. 8.12.

The triple-layer design fundamentally reimagines the traditional monolithic approach to panel construction. By layering the panel with a sturdier material at its core, the structural integrity is significantly bolstered. The middle layer, carefully selected for its superior strength properties, assumes a pivotal role in resisting bending forces, which are prevalent in various structural applications.

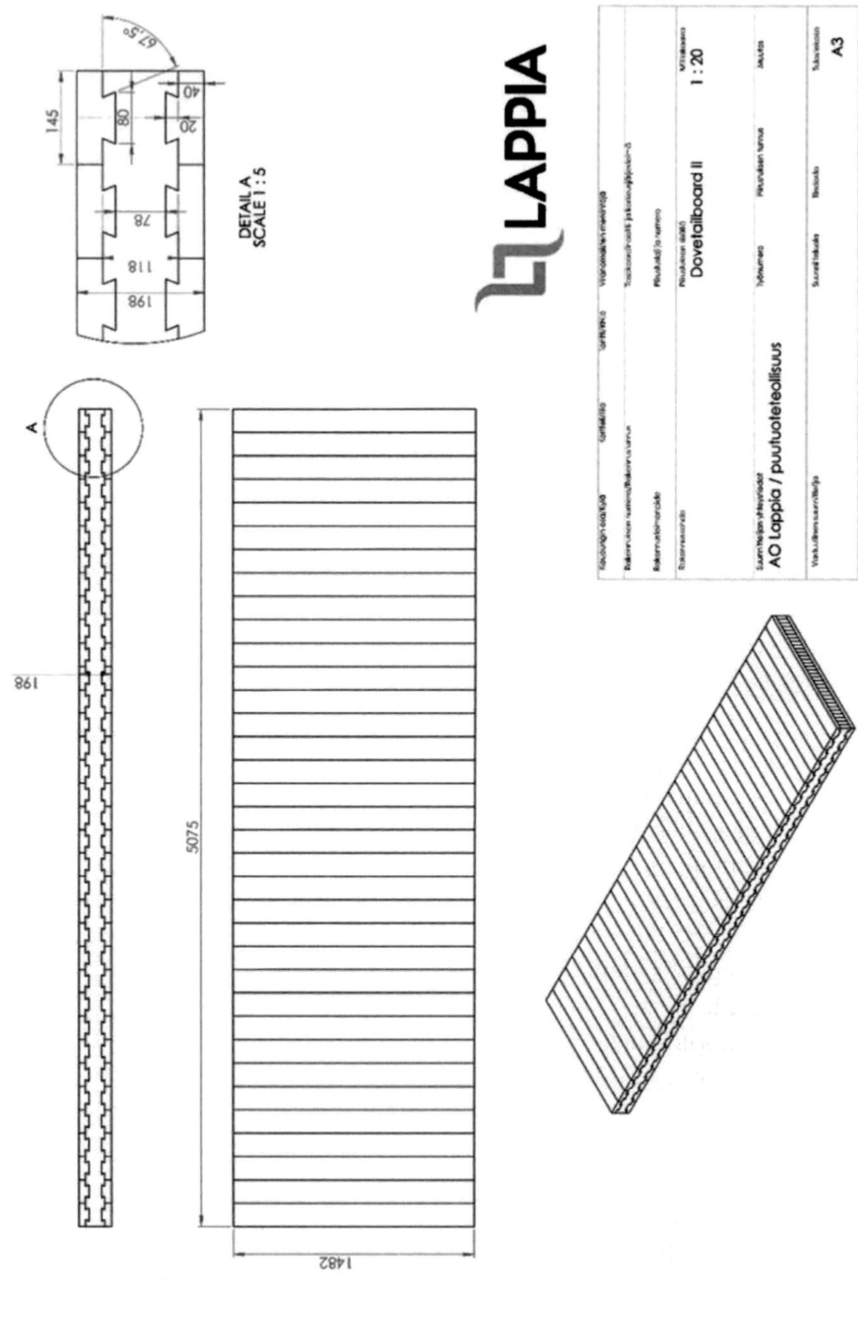
DETAIL A
SCALE 1 : 5
67,5°
145
80
40
20
78
118
198
A
198
5075
1482
LAPPIA
Dovetailboard II
1 : 20
AO Lappia / puutuoteteollisuus
A3

Figure 8.12 Dovetail massive wood board elements reinforced with 3-layer.

In Fig. 8.12, this concept is visually elucidated. The strategic placement of the robust middle layer ensures that it bears the brunt of bending stress, effectively mitigating the strain experienced by the panel. This layer acts as a core support, providing the necessary rigidity to counter bending forces and distribute them more evenly across the structure.

The outer layers of the triple-layer design, while not as robust as the middle layer, contribute to the overall structural stability and provide a protective encasement. They act as shields, defending the central layer against external pressures and environmental factors. This composite layering not only reinforces the panel's bending strength but also enhances its durability and longevity.

Furthermore, this triple-layer approach offers versatility in material selection, allowing for customization based on specific project requirements. Engineers and architects can tailor the composition of each layer to optimize the panel's performance, striking an ideal balance between strength, weight, and cost-effectiveness.

8.6 Concluding Remarks

In the contemporary timber market, the increasing demand for sustainable and eco-friendly building materials has sparked the development of various non-glue and non-metal wooden boards. However, within this evolving landscape, a significant gap is noticeable—a lack of dovetail-based components specifically designed for these sustainable solutions. This gap poses a challenge for conducting a thorough comparative analysis, hindering a comprehensive understanding of how our proposed designs measure up against existing alternatives in terms of similarities, differences, advantages, and disadvantages.

The absence of a dovetail-based component tailored to these sustainable materials raises questions about the extent to which traditional joinery methods are being adapted and optimized for contemporary environmentally conscious engineering. How might the integration of dovetail joinery, known for its strength and durability, enhance the performance and longevity of these sustainable wooden boards? Furthermore, what potential sustainability benefits could be unlocked by incorporating dovetail joinery, considering its minimal waste generation and potential for disassembly and reuse?

Additionally, exploring the limitations and challenges associated with implementing dovetail-based components in non-glue and non-metal wooden boards is crucial. Are there structural or logistical challenges that make dovetail joinery impractical or less effective in this context? Understanding these potential drawbacks is vital for ensuring that proposed design concepts are both innovative and feasible for application in environmentally conscious engineering.

In essence, this research aims to delve into the nuances of incorporating dovetail joinery into sustainable building materials, shedding light on the uncharted territory of integrating traditional craftsmanship into modern eco-friendly design principles. By doing so, we strive to provide a more nuanced perspective on the potential advancements and considerations in sustainable construction, ultimately contributing to a more holistic and informed approach to environmentally conscious engineering.

The integration of dovetail-based components in non-glue and non-metal wooden boards holds promising potential. Dovetail joinery is renowned for its durability, strength, and aesthetic appeal, making it a coveted choice in traditional woodworking. However, its application in the realm of sustainable, eco-conscious engineering has been overlooked until now. By introducing pioneering designs that incorporate dovetail joinery into wooden boards, we aim to unlock a new dimension of possibilities for environmentally friendly construction materials.

Moreover, our research intends to explore and elucidate the advantages of dovetail-based elements in wooden boards devoid of glue or metal fastenings. These advantages encompass enhanced structural integrity, increased load-bearing capacity, reduced reliance on environmentally harmful adhesives, and ease of disassembly for potential reuse or recycling. Moreover, dovetail-based designs align with sustainable principles, promoting the use of renewable resources and minimizing the carbon footprint associated with traditional construction methods.

In fostering a deeper understanding of these benefits and their implications, we endeavor to provide a comprehensive comparative analysis. This analysis will encompass various aspects, including manufacturing efficiency, cost-effectiveness, environmental impact, and overall performance when compared to existing alternatives. By conducting a thorough examination of the proposed DMWBEs and their integration into sustainable building practices, we aim to pave the way for a paradigm shift in the timber market towards more eco-friendly and socially responsible choices.

To initiate the integration of DMWBEs into the international building market and potentially replace EWPs, architectural modeling software was employed as the initial step. While the conclusions drawn from this investigation are primarily rooted in theoretical deliberations and have yet to undergo practical testing, they are proposed in accordance with established construction procedures. Subsequent steps involve evaluating the technical performance aspects of the developed designs, making necessary optimizations, and refining the inventions through market research.

Despite the limited current utilization of DMWBEs in commercial and structural contexts, recent investigations like The DoMWoB project at the Faculty of Built Environment, Tampere University, Finland, have shed light on promising prospects for the "innovative dovetail concept." This concept, inspired by one of the most ancient joining methods, holds immense potential for further development and broader utilization.

A crucial aspect is prioritizing further research and development on available materials and establishing a rigorous certification process to successfully introduce these innovative construction products into the market. Ensuring alignment with national design codes is pivotal for manufacturers, designers, and contractors to specify these products effectively. Comprehensive research and development efforts need to cover various aspects, including dimensional stability, response to vibration, long-term creep behavior, performance under outdoor conditions, thermal conductivity, and life-cycle impact analysis.

Furthermore, exploring other dovetail assembly techniques involving wood welding and wood nails is essential. These techniques could serve as inspiration for the potential application of sophisticated wood-modification technology and machining methods, revitalizing ancient wood assembly approaches. The amalgamation of these efforts will play a pivotal role in the advancement and integration of DMWBEs, potentially reshaping the landscape of sustainable and efficient construction practices.

Funding

This project has received funding of €202,680.96 from the European Union's Horizon 2020 research and innovation program under the Marie Skłodowska-Curie grant agreement no. 101024593.

This project has also received funding of €60,000 and €33,000 from the Marjatta and Eino Kolli Foundation for technical performance tests including fire safety, structural, air permeance, and sound insulation, and potential patent applications.

References

1. Stark, N.M. Chapter 11: Wood-Based Composite Materials-Panel Products- Glued-Laminated Timber, Structural Composite Lumber, and Wood–Nonwood Composite Materials. *Wood Handbook;* Department of Agriculture, Forest Service, Forest Products Laboratory: Madison, WI, USA, 2010, 190.
2. Gong, M. Engineered Wood Products for Construction. In *Engineered Wood Products for Construction*; Books on Demand; IntechOpen: Norderstedt, Germany, 2022.
3. Woodard, A.C.; Milner, H.R. Sustainability of Timber and Wood in Construction. In *Sustainability of Construction Materials*; Elsevier: Amsterdam, The Netherlands, 2016, 129–157.

4. Buck, D.; Wang, X.; Hagman, O.; Gustafsson, A. Comparison of Different Assembling Techniques Regarding Cost, Durability, and Ecology—A Survey of Multi-Layer Wooden Panel Assembly Load-Bearing Construction Elements. *BioResources* 2015, 10, 8378–8396.
5. Stepinac, M.; Šušteršic, I.; Gavric, I.; Rajcic, V. Seismic Design of Timber Buildings: Highlighted Challenges and Future Trends. *Appl. Sci.* 2020, 10, 1380.
6. Karjalainen, M. A Study of the Finnish Multi-story Timber Frame Buildings 1995–2018 S. ARCH 2019. In *Proceedings of the 6th International Conference on Architecture and Engineering*, Havana, Cuba, 5–7 March 2019.
7. Ramage, M.H.; Burridge, H.; Busse-Wicher, M.; Fereday, G.; Reynolds, T.; Shah, D.U.; Wu, G.; Yu, L.; Fleming, P.; Densley-Tingley, D.; et al. The Wood from the Trees: The Use of Timber in Construction. *Renew. Sustain. Energy Rev.* 2017, 68, 333–359.
8. Shahnewaz, M.; Tannert, T.; Alam, M.S. Popovski, M. In-Plane Stiffness of Cross-Laminated Timber Panels with Openings. *Struct. Eng. Int.* 2017, 27, 217–223.
9. Harte, A.M. Mass Timber-The Emergence of a Modern Construction Material. *J. Struct. Integr. Maint.* 2017, 2, 121–132.
10. Sikora, K.S.; McPolin, D.O.; Harte, A.M. Effects of the Thickness of Cross-Laminated Timber (CLT) Panels Made from Irish Sitka Spruce on Mechanical Performance in Bending and Shear. *Constr. Build. Mater.* 2016, 116, 141–150.
11. Abrahamsen, R.B. *Mjøstårnet-Construction of an 81 m Tall Timber Building*; Internationales Holzbau-Forum IHF: Garmisch-Partenkirchen, Germany, 2017.
12. CTBUH. Council on Tall Buildings and Urban Habitat. Illinois Institute of Technology, S.R. Crown Hall, 3360 South State Street, Chicago, Illinois, USA. Available online: http://www.ctbuh.org [Accessed: September 18, 2023].
13. HoHo. Available online: http://www.hoho-wien.at/ [Accessed: September 18, 2023].
14. Abrahamsen, R.B.; Malo, K.A. Structural Design and Assembly of "Treet"–A 14-Storey Timber Residential Building in Norway. In Proceedings of the WTCE 2014, *World Conference on Timber Engineering*, Quebec City, QC, Canada, 10–14 August 2014.
15. Puettmann, M.E.; Wilson, J.B. Gate-to-Gate Life-Cycle Inventory of Glued-Laminated Timbers Production. *Wood Fiber Sci.* 2005, 37, 99–113.

16. Cadorel, X.; Crawford, R. *Life Cycle Analysis of Cross Laminated Timber in Buildings: A Review.* The Architectural Science Association and RMIT University: Melbourne, Australia, 2018.
17. Burnard, M.; Tavzes, Č.; Tošić, A.; Brodnik, A.; Kutnar, A. The Role of Reverse Logistics in Recycling of Wood Products. *Environ. Footprints Eco-Design Prod. Process* 2015, 1–30.
18. Puettmann, M.; Sinha, A.; Ganguly, I. Life Cycle Energy and Environmental Impacts of Cross Laminated Timber Made with Coastal Douglas-Fir. *J. Green Build.* 2019, 14, 17–33.
19. Adhikari, S.; Ozarska, B. Minimizing Environmental Impacts of Timber Products through the Production Process "From Sawmill to Final Products". *Environ. Syst. Res.* 2018, 7, 6.
20. Lin, C.C.; Yu, K.P.; Zhao, P.; Whei-May Lee, G. Evaluation of Impact Factors on VOC Emissions and Concentrations from Wooden Flooring Based on Chamber Tests. Build. *Environ.* 2009, 44, 525–533.
21. Zelinka, S.L.; Sullivan, K.; Pei, S.; Ottum, N.; Bechle, N.J.; Rammer, D.R.; Hasburgh, L.E. Small Scale Tests on the Performance of Adhesives Used in Cross Laminated Timber (CLT) at Elevated Temperatures. *Int. J. Adhes. Adhes.* 2019, 95, 102436.
22. Peng, L.; Hadjisophocleous, G.; Mehaffey, J.; Mohammad, M. Fire Resistance Performance of Unprotected Woodwoodwood and Woodsteelwood Connections: A Literature Review and New Data Correlations. *Fire Saf. J.* 2010, 45, 392–399.
23. Emberley, R.; Inghelbrecht, A.; Yu, Z.; Torero, J.L. Self-Extinction of Timber. *Proc. Combust. Inst.* 2017, 36, 3055–3062.
24. Toivonen, R.; Lähtinen, K. *Sustainability—A Literature Review on Concealed Opportunities for Global Market Diffusion for the Crosslaminated Timber (CLT) in the Urbanizing Society*; BioProducts Business: Curitiba, Brazil, 2019.
25. Ilgın, H.E.; Karjalainen, M.; Koponen, O. *Review of the Current State-of-the-Art of Dovetail Massive Wood Elements*; IntechOpen: London, UK, 2021.
26. Sumiyoshi, T; Matsui, G. *Wood Joists in Classical Japanese Architecture,* Kajima Institute Publishing Co. Ltd., Japan, 1990.
27. Pang, S-J; Oh, J-K; Park, C-Y; Lee, J-J. Influence of Crossing-Beam Shoulder and Wood Species on Moment-Carrying Capacity in a Korean Traditional Dovetail Joint. *Journal of Wood Science* 2011, 57, 195–202.

28. Zhang, J.; Yixiang, Xu Y.; Mei, F.; Li, C. Experimental Study on the Fire Performance of Straight-Line Dovetail Joints. *Journal of Wood Science* 2018, 64, 193–208.
29. Zhang, J.; Wang, Y.; Linfeng, L.; Xu Q. Thermo-Mechanical Behaviour of Dovetail Timber Joints Under Fire Exposure. *Fire Safety Journal* 2019, 107, 75–88.
30. Yingzao Fashi, Q.G. Twelfth-Century Chinese Building Manual. *Archit. Hist.* 1998, 41, 1–13.
31. Rivers, S; Umney, N. *Conservation of Furniture*, Oxford, England: Routledge, 2003.
32. *Routledge French Technical Dictionary*. 1. London: Routledge, 1994. (ISBN 9780415112253).
33. Jasieńko, J.; Nowak, T.; Karolak, A. Historical Carpentry Joints. *Journal of Heritage Conservation* 2014, 40.
34. Kopkowicz, F. *Ciesielstwo Polskie*, Arkady, Warszawa, 1958.
35. Tannert, T.; Keller, N.; Frei, R.; Vallee T. Improved Performance of Rounded Dovetail Joists, WTCE, *World Conference on Timber Engineering*, July 15–19, Auckland, New Zealand, 2012.
36. Pang, S-J; Oh, J-K; Park, C-Y; Lee, J-J. Effects of Size Ratios on Dovetail Joints in Korean Traditional Wooden Building, WTCE, *World Conference on Timber Engineering*, July 15–19, Auckland, New Zealand, 2012.
37. Rinne, R.; Ilgın, H.E.; Karjalainen, M. Comparative Study on Life-Cycle Assessment and Carbon Footprint of Hybrid, Concrete and Timber Apartment Buildings in Finland. *Int. J. Environ. Res. Public Health* 2022, 19, 774.
38. Tulonen, L.; Karjalainen, M.; Ilgın, H.E. *Tall Wooden Residential Buildings in Finland: What Are the Key Factors for Design and Implementation?* IntechOpen: London, UK, 2021.
39. Soikkeli, A.; Ilgın, H.E.; Karjalainen, M. Wooden Additional Floor in Finland. *Encyclopedia* 2022, 2, 578–592.
40. Karjalainen, M.; Ilgın, H.E. A Statistical Study on Multi-story Timber Residential Buildings (1995–2020) in Finland. In *Proceedings of the LIVENARCH VII Livable Environments & Architecture 7th International Congress OTHER ARCHITECT/URE(S)*, Trabzon, Turkey, 28–30 September 2021, 1, 82–94.
41. Cabral, J.P.; Kafle, B.; Subhani, M.; Reiner, J.; Ashraf, M. Densification of Timber: A Review on the Process, Material Properties, and Application. *J. Wood Sci.* 2022, 68, 20.

42. Xu, B.; Liu, K.; Zhao, Y.; Bouchaïr, A. Pullout Resistance of Densified Wood Dowel Welded by Rotation Friction. *J. Mater. Civ. Eng.* 2022, 34, 04022186.

43. Pereira, M.C.d.M.; Pascal Sohier, L.A.; Descamps, T.; Junior, C.C. Doweled Cross Laminated Timber: Experimental and Analytical Study. *Constr. Build. Mater.* 2021, 273, 121820.

44. Bell, T. A Detailed Investigation into the Engineering Properties and Challenges Affecting the Potential Introduction of a UK Grown Dowel-Laminated Timber Floor Panel into the Domestic Construction Market. Ph.D. Thesis, University of Strathclyde, Glasgow, Scotland, 2018.

45. Xu, Y.; Dong, Z.; Jia, C.; Wang, Z.; Lu, X. Bamboo Nail: A Novel Connector for Timber Assemblies. *J. Renew. Mater.* 2021, 9, 1609–1620.

46. Rezaei Rad, A.; Burton, H.; Rogeau, N.; Vestartas, P.;Weinand, Y. A Framework to Automate the Design of Digitally-Fabricated Timber Plate Structures. *Comput. Struct.* 2021, 244, 106456.

47. Drdácký, M.; Urushadze, S. Retrofitting of Imperfect Halved Dovetail Carpentry Joints for Increased Seismic Resistance. *Buildings*, 2019, 9, 48. doi:10.3390/buildings9020048.

48. Xie, Q.; Zhang, B.; Zhang, L.; Guo, L.;Wu, Y. Normal Contact Performance of Mortise and Tenon Joint: Theoretical Analysis and Numerical Simulation. *J. Wood Sci.* 2021, 67, 31.

49. Gamerro, J.; Bocquet, J.F.; Weinand, Y. A Calculation Method for Interconnected Timber Elements Using Wood-Wood Connections. *Buildings* 2020, 10, 61.

50. Sha, B.; Wang, H.; Li, A. The Influence of the Damage of Mortise-Tenon Joint on the Cyclic Performance of the Traditional Chinese Timber Frame. *Appl. Sci.* 2019, 9, 3429.

51. Jeong, G.Y.; Song, J.K. Evaluation of Structural Properties of Dovetail Connections under Tensile Load Using Three Methods of Data Analysis. *J. Mater. Civ. Eng.* 2017, 29, 06017011.

52. Branco, J.M.; Descamps, T. Analysis and Strengthening of Carpentry Joints. *Constr. Build. Mater.* 2015, 97, 34–47.

53. Jeong, G.Y.; Park, M.J.; Park, J.S.; Hwang, K.H. Predicting Load-Carrying Capacity of Dovetail Connections Using the Stochastic Finite Element Method. *Wood Fiber Sci.* 2012, 44, 430–439.

54. Ozkaya, K.; Burdurlu, E.; Ilce, C.; Ciritcioglu, H.H. Diagonal Tensile Strength of An Oriented Strandboard (OSB) Frame with Dovetail Corner Joint. *BioResources* 2010, 5, 2690–2701.

55. Jeong, G.Y.; Park, M.-J.; Hwang, K.; Park, J.S. Effects of Geometry on Mechanical Behavior of Dovetail Connection, WTCE. In *Proceedings of theWorld Conference on Timber Engineering*, Auckland, New Zealand, 15–19 July 2012.
56. Robeller, C.; Weinand, Y. Interlocking Folded Plate-Integral Mechanical Attachment for Structural Wood Panels. *Int. J. Space Struct.* 2015, 30, 111–122.
57. Pozza, L.; Scotta, R.; Trutalli, D.; Polastri, A. Behaviour Factor for İnnovative Massive Timber Shear Walls. *Bull Earthq. Eng.* 2015, 13, 3449–3469.
58. Fraser, M. *Design Research in Architecture: An Overview*, London: Ashgate, 2013.
59. Akšamija, A. *Research Methods for the Architectural Profession*, New York, NY, USA: Routledge, 2021.
60. Short, C.A. What İs "Architectural Design Research"? *Building Research & Information* 2008, 36(2), 195–199. DOI: 10.1080/09613210701811015.
61. Vasilenko, N.A. General System Principles of Architectural Systems Formation. *IOP Conference Series: Materials Science and Engineering* 2020, 753, 32047.
62. Luck, R. Design Research, Architectural Research, Architectural Design Research: An Argument on Disciplinarity and İdentity. *Design Studies* 2019, 65, 152–166. DOI: 10.1016/j.destud.2019.11.001.
63. Gao, Y. Design Research in Architecture: An Overview, Edited by Murray Fraser. *The Design Journal.* 2015, 18(2), 291–294. DOI: 10.2752/17563 0615X14212498964439.
64. Eilouti, B. Scenario-Based Design: New Applications in Metamorphic Architecture. *Frontiers of Architectural Research* 2018, 7(4), 530–543. DOI: 10.1016/j.foar.2018.07.003.
65. Luck, R. Participatory Design in Architectural Practice: Changing Practices in Future Making in Uncertain Times. *Design Studies* 2018, 59, 139–157. DOI: 10.1016/j.destud.2018.10.003.
66. Bhooshan, S. Parametric Design Thinking: A Case-Study of Practice-embedded Architectural Research. *Design Studies* 2017, 52, 115–143. DOI: 10.1016/j.destud.2017.05.003.
67. Fu, F. *Design and Analysis of Tall and Complex Structures*. Oxford and Cambridge: Butterworth-Heinemann, Elsevier, 2018.

Chapter 9

Performance Assessment of Electric-Powered Food Roasting Machine

Joseph Abutu,[a] Emeka Daniel Oruonye,[b] Mohammed Abubakar,[a] and Abdullahi Auwal[a]

[a]*Department of Mechanical Engineering, Taraba State University, Jalingo, Nigeria*
[b]*Department of Geography, Taraba State University, Jalingo, Nigeria*

eoruonye@gmail.com

Over the years, traditional roasting systems in which foods are roasted manually over an open fire utilizing biomass fuels, such as wood, twigs, dung, or coal, have been utilized for roasting agricultural produce. Hence, in this study, an electric-powered food roasting machine that utilizes electrical energy was developed using locally available materials with the aim of investigating the performance of the constructed oven using standard test procedures. The performance of the constructed baking system was assessed by investigating the moisture content and roasting rate. Two different samples were utilized during the testing process (maize and chicken) and roasting samples were carried out for 2100 seconds. The experimental results revealed that the roasting rate (1.947 g/min) was the highest at 86.9 °C during the roasting of chicken while the roasting rate (0.6958 g/min)

Reclaiming Eden: Responsible Living, Engineering, and Architectures
Edited by David S.-K. Ting and Jacqueline A. Stagner

ISBN 978-981-5129-31-1 (Hardcover), 978-1-003-49612-0 (eBook)
www.jennystanford.com

was the highest at 99 °C during the roasting of maize. In addition, moisture contents were the lowest (0.431%) and the highest (2.29%) at 218.3 °C and 86.9 °C, respectively, during the roasting of chicken while moisture contents were the lowest (0.808%) and the highest (4.761%) at 218.3 °C and 99 °C, respectively, during roasting of maize. Also, the total cost of developing the electric-powered roasting machine is lower compared to the cost of buying a new electric-powered roasting machine in the market. Hence, it can be concluded that the developed roasting machine is cost-effective and capable of enhancing an efficient and effective roasting process. However, the roasting time depends on the weight of the test samples.

9.1 Introduction

The success and growth of any industry depend on the quality of products and energy-efficient processes. However, food processing techniques, such as roasting, frying, boiling, and baking, are some of the most energy-intensive unit operations (Mukherjee et al., 2017). These processing techniques involve the use of engineering technology in conjunction with food science to transform or alter the look, taste, scent, shape, or color of the food (Zhang, 2019). Wang et al. (2016) revealed that there are numerous food processing procedures available, the majority of which are used in food industries. Though, heat processing, size reduction, and mixing are some of these strategies (Wang et al., 2016). Nevertheless, baking and roasting are considered common cooking techniques comprising heating food inside an oven at a constant temperature. Hence, the heat is transferred to the load solely by convection and radiation (Adegbola et al., 2018). Roasting is a processing method that alters the food quality through heated air to cause some desirable chemical and physical changes and to enhance food flavor (Sruthi et al., 2021). The work of Luigi et al. (2017) has found that the characteristic flavor of food is developed during roasting while Makky and Cherie (2021) reported that roasting causes complex changes to the component of the food at the surface. These changes retain moisture in the bulk of the food and enhance eating quality. Hence, these desirable qualities developed in food products are factors that enhance their use as raw materials for large- and small-scale processors and also increase the market value of the products (Sudirman 2022). In addition, Samuel et al. (2022) revealed that the secondary purpose of roasting is the reduction of water activity at the surface of the food and the preservation by destruction of microorganisms.

Traditionally, foods are roasted on an open fire in various pots such as clay, aluminum, or iron pots (Kabri, 2011). Though, in some conventional food roasting methods, food is roasted by placing charcoal within a tray and a metal grill on the top of a tray, so as to separate the food to be roasted from the charcoal (Awopetu and Aderibigbe, 2017). These roasting techniques, most times, are laborious as an operator gets tired of waiting for the food to be completely roasted (Ross, 2018). In addition, direct touch and exposure of the human body to direct heat released by charcoal may have an adverse effect on human health (Awopetu and Aderibigbe, 2017). The report of Zhang et al. (2021) affirmed that the roasting of foods using these traditional methods increases the antioxidant content, lowers their sugar level, and creates volatile chemicals that improve flavor. Hence, there is a need for researchers to develop an alternative indigenous roasting machine, which is efficient, safe to use, and can be applied for commercial purposes. However, several authors have developed food-roasting machines using different sources of energy. Some of these works include the work of Idowu et al. (2020), where an electrically operated groundnut roasting machine was developed and compared with traditional methods. The authors found that the traditional and new methods recorded an average roasting efficiency of 83.85% and 94.11%, respectively. Also, Fachruddin et al. (2021) developed an electric-powered coffee roasting machine and reported that a roasting machine produces an average temperature of 196.64 °C with a final moisture content of 3.61% in 50 mins. In addition, Awode et al. (2023) developed a dual-powered (electric and gas) food baking oven and revealed that the length of roasting has a big impact on how strong the flavor and aroma. The authors also revealed that more heat was expended when baking with electricity (39.742 kJ) than when baking with gas (0.334 kJ). Consequently, the need to regulate the amount of heat that enters the heating chamber using a temperature regulator of high-temperature range has not been fully explored. Hence, in this study, an electric-powered roasting machine with a digital temperature control box (0–300 °C) has been developed using a scrapped microwave system by incorporating an electric heating element into the interior with the need to promote the development of high-performing indigenous roasting machines, which incorporate a temperature regulator.

9.2 Materials and Methods

9.2.1 Materials

The material selected for this study includes a locally sourced scraped microwave oven casing that serves as the roasting chamber or internal walls, mild steel (1 mm metal sheet), and fiberglass that serves as a lagging material. Other materials used in this work include a 1.5 mm pipe for constructing the frame, a 1 mm angle iron for fastening the control panel on the frame, a panel, and four wheels to enable the oven to move freely from one point to another. The test equipment utilized in the work includes an infrared thermometer [Model: GM320-BENETECH, measurement range: –50–4000 °C; accuracy: –50 °C ~ 0 °C (±3 °C), 0 °C ~400 °C (±1.5 °C] and stopwatch. All materials were locally sourced from various commercial shops in Jalingo, Nigeria. In addition, a microwave heating system was used as a heating element (1800 W). This decision was based on the findings of Kamble (2015), who revealed that microwave pyrolysis does not only address the drawbacks of traditional pyrolysis technologies (slow heating and the need for feedstock shredding) but also increases the quality of the final pyrolysis products.

9.2.2 Methods

9.2.2.1 Design consideration

Several factors were considered during the construction of the food roaster. These factors include ease of size of an oven, assembly of the oven parts, size of chicken or maize, roasting temperature, and time taken to roast maize and chicken. Also, in order to achieve high reliability, efficiency, and cheaper methods of developing a food roasting machine, principles of operation of the roaster, convenience of operation, rate of heat transfer during roasting to meet the needs, and ease of producing the system were considered as a design consideration.

9.2.2.2 Design drawings

AutoCAD 2020 was used for design drawings. The orthographic design of the food roasting machine is shown in Fig. 9.1 while the assembly drawing (isometric view) is shown in Fig. 9.2.

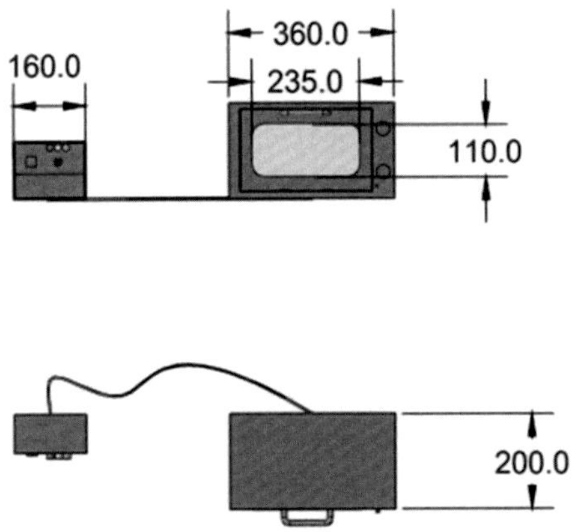

Figure 9.1 The orthographic view of the food roasting machine (dimension in mm, 1:25).

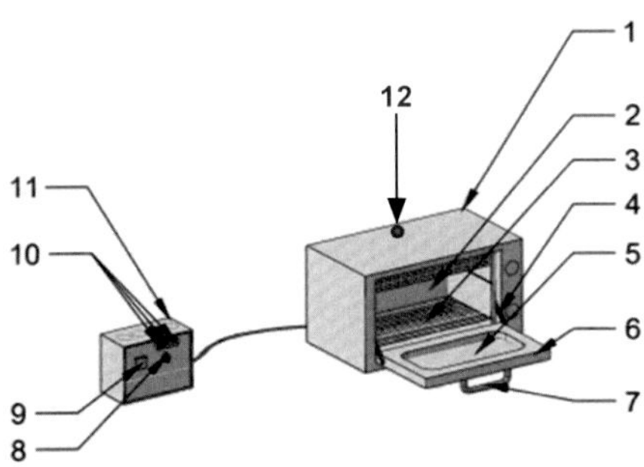

List of components	
1	Roasting Oven
2	Chamber
3	Tray
4	Linkage
5	Viewing glass
6	Cover
7	Handle
8	Control Knob
9	Display
10	Indicator lights
11	Power control unit
12	Vent

Figure 9.2 Assembly drawing (isometric view).

9.2.2.3 Fabrication and assembling of parts

The fabrication of the framework was initiated by constructing the frame using locally sourced mild steel, which was cut to the desired size, as stipulated in Fig. 9.1, and welded together, making provisions for the installation of other key components of the roasting machine. This fabrication was carried out at the Mechanical Engineering workshop of Taraba State University. As recommended by Awode et al. (2023), the components were assembled and squared using a try-square before being welded with a filler metal, an electrode, as well as an arc welding machine.

(a)

(b)

Figure 9.3 Constructed food roasting machine: (a) Outer view and (b) Inner view of the roasting machine.

In addition, the roasting chamber or internal wall was thereafter riveted, and two chrome coils were attached to the top and bottom sides of the internal walls. Afterward, the fiberglass was loaded, and the viewing glass and door handle were secured. In order to aid the movement of the machine from one location to another, rollers were fixed on each stand of the machine. These rollers will help protect the oven from direct surface contact, avoiding corrosion and surface scratches caused by friction. Thereafter, a vent (20 mm) was introduced at the top center for the continuous removal of the hot and humid air from the inner baking chamber during roasting (Vogel 2005). The thermostat (0–300 °C), power switch, and indicator light were fitted using nuts and bolts. To eliminate contours and irregular areas on the body of the roaster, a grinding machine was used and thereafter, painted. The operating temperature of the machine is 0–300 °C. Figure 9.3 shows the assembled parts of the food roasting machine.

9.2.2.4 Performance evaluation

The performance evaluation of the developed food roasting machine was carried out using fresh maize and chicken. The maximum heating temperature (220–300 °C) of the roaster was obtained in 16 mins. Also, in order to get the calibrated mark of the thermometer, the maximum temperature of the oven was divided into six (50, 100, 150, 200, 250, and 300 °C). However, before calibrations were made, it was found that the external body of the oven was about 41 °C when the internal environment was 227 °C. Hence, several efforts were made in order to minimize the external body temperature while maximizing the internal temperature (Awode et al., 2023). This was done by disassembling the oven and separating the internal connection from the external using wood (insulator). Thereafter, the oven was assembled, and the experiment was performed at time intervals of 5 mins using three specimens each from maize and chicken parts. Hence, experiments were conducted for each of the calibrated thermostats while measurements were taken with respect to the corresponding time taken and temperature for a particular level of the thermostat knob. For each test sample, an experiment was conducted by recording the weight of the test sample and the temperature of the roasting chamber. The roasting of the test sample is shown in Fig. 9.4. In addition, the results obtained were used to calculate the moisture content (Eq. (9.1)) and roasting rate (Eq. (9.2)) of test samples.

Figure 9.4 Experimental test samples.

$$\%\ \text{Moisture content} = \frac{m_o - m_r}{m_o} \times 100 \tag{9.1}$$

$$\text{Roasting rate} = \frac{m_o - m_r}{t} \tag{9.2}$$

where, m_o = mass of sample before roasting (g), m_r = mass of roasted sample (g), and t = roasting time (min).

Also, the quantity of heat required (Q) to roast the test samples was calculated using Eq. (9.3).

$$Q = m_r \times c_h \times \Delta T \tag{9.3}$$

$$\Delta T = T_2 - T_1 \quad (9.4)$$

Where, ΔT = change in roasting temperature, c_h = specific heat capacity of the test sample, T_2 = final roasting temperature, and T_1 = initial roasting temperature.

According to TET (2023), the specific heat capacity for fresh chicken (broiler), c_h = 3.22 kJ/kg°C and c_h for maize = 2.99 kJ/kg°C.

9.3 Results and Discussion

9.3.1 Experimental Results

The experimental results for the performance assessment of the developed food roaster using maize and chicken are presented in Tables 9.1 and 9.2. The highest heat supplied (167.04 kJ) was obtained in the 35th minute during the roasting of chicken while the lowest heat supplied (1.21 kJ) was also achieved in the 35th minute during the roasting of maize. These results are represented in Fig. 9.5. Based on the results, it can be revealed that more heat was supplied to the roasting machine during the roasting of chicken compared to that of maize. This may be attributed to the weight of the chicken sample. Awode et al. (2023) have revealed that excess weight has a significant adverse impact on natural cooling mechanisms. In addition, Fig. 9.6 shows that the lowest moisture content (0.808%) was recorded during the roasting of maize at the 35th minute. These results showed an improvement compared with the previous work of Fachruddin et al. (2021), who developed an electric-powered roasting machine with a similar heating element capacity (1800 W) and reported a moisture content of 3.61% using an agro-based product (coffee) as a test sample. This improvement may be attributed to the earlier report of Guo et al. (2021), who revealed that water content decreases at high temperatures because higher temperature leads to higher energy, which moves to the material to accelerate the evaporation of water. Also, Fig. 9.7 reveals that the highest roasting rate (5.84 g/min) was achieved during chicken roasting with the highest value achieved at the 10th minute. This may be attributed to the weight differences between the two experimental samples.

Table 9.1 Experimental results for chicken roasting (initial mass (m_o) = 1304.4 g)

Run	Time (min)	Sample mass, m_r (g)	Initial roasting temperature, T_1 (°C)	Final roasting temperature, T_2 (°C)	ΔT (°C)	Quantity of heat (kJ)	Moisture content (%)	Roasting rate (g/min)
1	10	1275.2	62	86.9	24.9	102.24	2.290	5.84
2	15	1257	86.9	111.8	24.9	100.78	1.427	3.64
3	20	1244.5	111.8	134.2	22.4	89.76	0.994	2.5
4	25	1234.7	134.2	153.7	19.5	77.53	0.787	1.96
5	30	1228.8	153.7	175.9	22.2	87.84	0.478	1.18
6	35	1223.5	175.9	218.3	42.4	167.04	0.431	1.06

Table 9.2 Experimental results for maize roasting (initial mass (m_o) = 153.16 g)

Run	Time (min)	Sample mass, m_r (g)	Initial roasting temperature, T_1 (°C)	Final roasting temperature, T_2 (°C)	ΔT (°C)	Quantity of heat (kJ)	Moisture content (%)	Roasting rate (g/min)
1	10	146.2	92	99	7	3.06	4.761	1.392
2	15	140.9	101	108	7	2.95	3.625	1.06
3	20	139.1	109	114	5	2.08	1.278	0.36
4	25	137.5	146	150	4	1.64	1.150	0.32
5	30	136.2	153	156	3	1.32	0.945	0.26
6	35	135.1	158	161	3	1.21	0.808	0.22

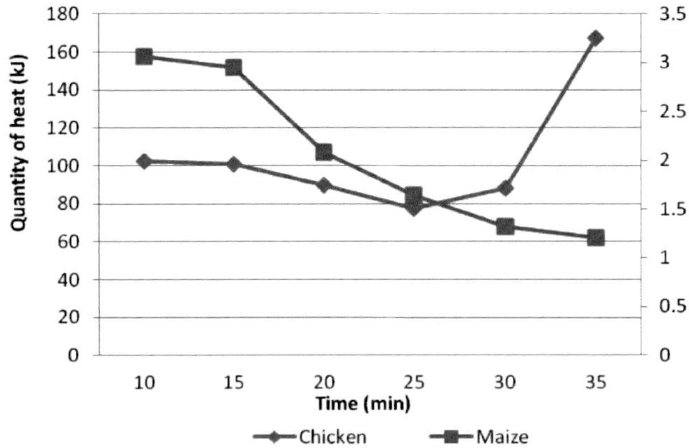

Figure 9.5 Quantity of heat supplied.

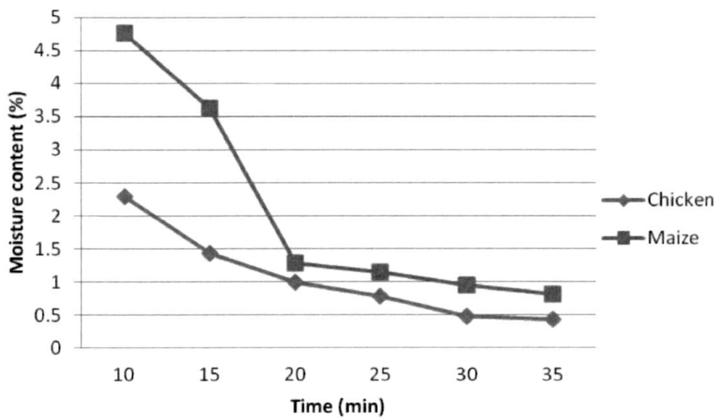

Figure 9.6 Moisture content in (a) baking bread and (b) roasting fish.

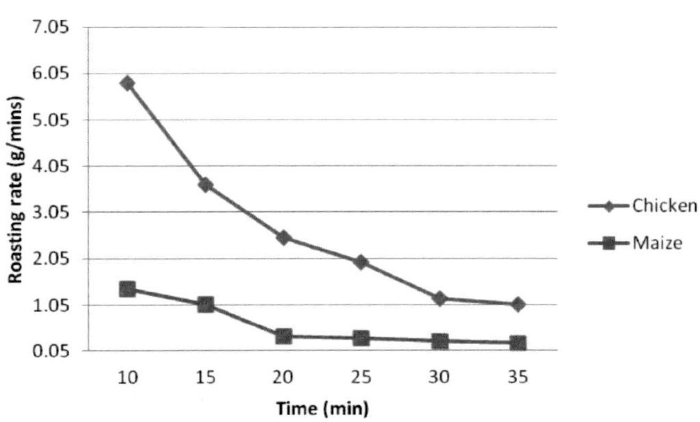

Figure 9.7 Roasting rate in (a) baking bread and (b) roasting fish.

9.3.2 Cost Analysis

The cost analysis of developing the roaster involves the preparation of cost data, which plays an important role in ensuring that all elements of the production cost are considered. The factors taken into consideration during the construction and production of the roaster include cost of materials (TCM), labor cost (TLC), and indirect or overhead cost (TOC). In this work, TLC and TOC were calculated using Eq. (9.5) and Eq. (9.6), respectively, while the total cost of production (TCP) was calculated using Eq. (9.7). The material costs (TCM) are presented in Table 9.3 while TLC involved the cost of fabrication of metallic parts (FC), cost of painting (CP), cost of transportation (CT) and the cost of installation of temperature regulator and other electrical parts (CE). Overhead cost (TOC) is assumed to be 30% of the TCM and TLC (Dongoyaro et al., 2021).

Table 9.3 Cost of materials in naira

S/N	Material	Quantity	Size	Unit	Amount (₦)
1	Scraped microwave oven	1	—	1	3,500
2	Digital temperature control box	1	—	1	30,000
3	Mild steel sheet	1	2 mm	—	2,000
4	Fiberglass	4 kg	10 mm	—	2,000
5	Square pipe	1	½ inch	1	1,500
6	Angle iron	1	½ inch	1	1,500
7	Quarters pipe	1	½ inch	1	1,200
8	Angle iron bar	1	1 inch	1	1,700
9	Body filler	—	—	—	1000
10	Electric cable	yard	3 mm	—	850
11	Paint	—	—	—	4,000
Total cost of materials (TCM)					**₦49,250**

Labor cost (TLC)

$$TLC = FC + CP + CT + CE \tag{9.5}$$

where FC = ₦6,000, CP = ₦3,000, CT = ₦2,000, and CE = ₦4,500.

TLC = ₦15,500

Overhead cost (TOC)

$$\mathrm{TOC} = \frac{\mathrm{TCM} + \mathrm{TLC}}{100} \times 30 \quad (9.6)$$

TOC = ₦19,425

Total cost of production (TCP)

$$\mathrm{TCP} = \mathrm{TLC} + \mathrm{TCM} + \mathrm{TOC} \quad (9.7)$$

TCP = 49,250 + 15,500 + 19,425 = ₦84,175

The cost of a baking oven in the market (Century 37L; SKU: CE125HL0T54F9NAFAMZ, Temp. 100–300 °C) = ₦91,800 (Source: Jumia.com, 2023).

These results indicate that the total cost of producing the electric-powered roasting machine (₦84,175) is lower compared to the cost of buying a new electric-powered roasting machine (₦91,800) of similar capacity in the market. Hence, it is cost-effective and recommended for use by low-income earners.

9.4 Conclusion

In this study, an electric-powered roasting machine was developed from scraps and locally sourced materials, and the performance was evaluated. Based on the results obtained, the following conclusions can be drawn:

i. Heat supplied increases with time. Hence, more heat was supplied during the roasting of chicken while the lowest heat supplied was achieved during the roasting of maize.
ii. The weight of test samples has a significant adverse impact on the quantity of heat supplied to the sample during roasting.
iii. The highest moisture content (4.761%) was recorded during maize roasting while the highest roasting rate (5.84 g/min) was achieved during chicken roasting.

iv. The mass of the test samples reduces as roasting time increases and vice versa.

v. The developed roasting machine is cost-effective and recommended for use by low-income earners.

References

Adegbola, A., Okoye, O., Adeyemo, T., Lameed, M., & Oke, H. (2018). Development and performance of baking oven. *Journal of Advanced in Research*, 17(3), 1–15.

Awode, E. I., Abutu, J., & Garkuwa, H. (2023). Performance assessment of dual-powered baking oven developed from locally sourced materials. *Journal of Advanced Industrial Technology and Application*, 4(1), 38–45.

Awopetu, O. O., & Aderibigbe, A. F. (2017). Development of a manually operated multi-purpose roasting machine. *Br J Appl Sci Technol*, 20(1), 1–7.

Dogonyaro, L. G., Okechukwu, C., Abutu, J., Akinode, R. A., Waino, A. J., & Adamu, E. (2021). Development of an agro-products dryer using locally sourced materials. *Taraba Journal of Engineering and Technology (TAJET)*, Faculty of Engineering, Taraba State University, Jalingo, 2(1), 56–61.

Fachruddin, F., Lubis, A., Maulina, H., & Nazura, P. (2021). Testing coffee roasting machine with electric heater as energy source. In *IOP Conference Series: Earth and Environmental Science*, 922(1), 012073. IOP Publishing.

Guo, X., Ho, C. T., Schwab, W., & Wan, X. (2021). Effect of the roasting degree on flavor quality of large-leaf yellow tea. *Food Chem.*, 347, 129016.

Idowu, A. A., Abigael, A. O., Kazeem, A. B., & Beatrice, O. M. (2020). Design, fabrication and evaluation of electrically-operated groundnut roasting machine. *American Journal of Engineering and Technology Management*, 5(3), 48–55.

Kamble, Y., & Patil, S. (2015). An overview of study of microwave heating technique and its applications. *IJSRD-International Journal for Scientific Research & Development*, 3(02), 2321–0613.

Luigi, P., Imre, B., Andreas, D., & Thomas, H. (2017). Chapter 12: The Chemistry of Roasting—Decoding Flavor Formation, Editor(s): Britta Folmer. *The Craft and Science of Coffee*, Academic Press, 273–309.

Makky, M., & Cherie, D. (2021). Pre-harvest oil palm FFB nondestructive evaluation technique using thermal-imaging device. *IOP Conference Series: Earth and Environmental Science*, 757(1), 012003.

Mukherjee, S., Asthana, A., Howarth, M., & Mcniell, R. (2017). Waste heat recovery from industrial baking ovens. *Energy Procedia*, 123, 321–328.

Ross, A. S. (2018). Flour quality and artisan bread. *Cereal Food World*, 63(2). Retried from https://doi.org/10.1094/CFW-63-2-0056.

Samuel, C. A., Kelvin, A., Kyle, N., & Reza T. (2022). Microstructural and physicochemical changes of coated and frozen fried chicken. *Journal of Food Processing and Preservation*, 10.1111/jfpp.16822, 46, 9.

Sruthi, N. U., Premjit, Y., Pandiselvam, R., Kothakota, A., & Ramesh, S. V. (2021). An overview of conventional and emerging techniques of roasting: Effect on food bioactive signatures. *Food Chemistry*, 348, 129088.

Sudirman, Z. (2022). The market turbulence, entrepreneurial marketing and marketing performance: An empirical study of tuna fillets SME's industry in Southeast Sulawesi, Indonesia. *Journal of Eastern European and Central Asian Research (JEECAR)*, 9(6), 1008–1020.

The Engineering Toolbox (2023). Food and foodstuff-specific heat. Retrieved from www.engineeringtoolbox.com on 15th May 2023.

Vogel A. (2005). *Textbook of Practical Organic Chemistry*, 3rd Edition.

Wang, C. Y., Huang, H. W., Hsu, C. P., & Yang, B. B. (2016). Recent advances in food processing using high hydrostatic pressure technology. *Critical Reviews in Food Science and Nutrition*, 56(4), 527–540.

Zhang, Y., Li, X., Lu, X., Sun, H., & Wang, F. (2021). Effect of oilseed roasting on the quality, flavor and safety of oil: A comprehensive review. *Food Research International*, 150, 110791.

Zhang, Z. H., Wang, L. H., Zeng, X. A., Han, Z., & Brennan, C. S. (2019). Non-thermal technologies and their current and future application in the food industry: A review. *International Journal of Food Science & Technology*, 54(1), 1–13.

Chapter 10

Coral Reef Ecosystem Enhancement in Singapore's Highly Urbanized Port

Chin Soon Lionel Ng,[a] Shu Qin Sam,[a] Yuichi Preslie Kikuzawa,[b] Tai Chong Toh,[a,c] and Loke Ming Chou[a]

[a]*Tropical Marine Science Institute, National University of Singapore, Singapore*
[b]*Centre for Liveable Cities, Ministry of National Development, Singapore*
[c]*College of Alice & Peter Tan, National University of Singapore, Singapore*

lionel.ngcs@gmail.com

Coastal development activities to support economic development and population growth have caused widespread degradation of Singapore's coral reefs. This provided opportunities for the development of science-based strategies to mitigate complete reef loss and ensure the maintenance of genetic diversity from affected areas. Here, we summarize key lessons from a five-year research program that aimed to maximize the conservation and restoration of stony coral diversity from Sultan Shoal, a southern offshore island predicted to be impacted by the expansion of Singapore's port operations. Prior assessments of environmental conditions helped determine the suitability of potential recipient sites as nurseries for stocking coral material from Sultan Shoal and subsequent transplant locations. The customization of nursery design and regular maintenance regimes were

Reclaiming Eden: Responsible Living, Engineering, and Architectures
Edited by David S.-K. Ting and Jacqueline A. Stagner

ISBN 978-981-5129-31-1 (Hardcover), 978-1-003-49612-0 (eBook)
www.jennystanford.com

essential for enhancing the survival and growth of transplanted corals. The suitability of 30 coral species as candidates for transplantation was tested, thereby generating growth rates for the planning of future restoration efforts. Concurrently, protocols for transplantation were also refined to achieve greater ecologically-effective outcomes. These efforts led to the provision of valuable ecosystem services to recipient sites, such as increased habitat for various reef biota and the contribution of genetic material from sexually mature coral transplants, underscoring the importance of regular monitoring to assess long-term outcomes of restoration.

10.1 Introduction

Coral reefs provide key functions such as food and nursery grounds that sustain at least 25% of the world's marine species and contribute ecosystem services including coastal protection and livelihoods for hundreds of millions of people [1]. However, significant reef degradation and hard coral loss driven largely by frequent and severe global-scale bleaching events have occurred particularly over the last three decades [1]. These trends are exacerbated by impacts from a multitude of localized anthropogenic activities such as coastal urbanization, pollution, and unsustainable resource use practices [2–4], which have resulted in far-reaching consequences. In particular, human pressures such as coastal development and overfishing have resulted in potentially irreversible losses in coral cover in Bahrain and China [5, 6]. A concomitant reduction of ecosystem functions has also been observed with reef degradation - some impacts include a reduction in the functional diversity of reef fishes with increased tourist developments in Mexico [7], and a decrease in coral carbonate production rates at sites nearer the highly urbanized Singapore mainland [8].

Ensuring that coral cover and diversity do not deteriorate further is therefore highly essential for maintaining the resilience and functioning of reef ecosystems. While coastal development projects often have to proceed for the purposes of economic growth but usually result in the destruction of large swathes of coral reef, mitigation strategies such as the planned relocation of organisms to other unimpacted sites can be carried out to reduce permanent losses in genetic diversity and reef function. This approach has been increasingly adopted at various scales, with different reef organisms such as hard corals, sponges, and echinoderms (e.g. [9–13]). These relocation efforts also provide valuable opportunities to establish new concepts and principles for coastal environment management and investigate scientific aspects to fill knowledge gaps.

The need to develop sustainably is especially important in tropical Southeast Asia, where almost all coastal cities are associated with coastal developments located in close proximity to coral reefs and marine protected areas. Singapore, for example, has seen significant changes to its marine and coastal environment over the past five decades. Its territorial waters are among the world's busiest, with more than 80% designated for port and shipping activities, while the remaining areas are used by various sectors such as military, petrochemical, aquaculture, and recreation [14]. To cater to economic development and population expansion, land area has increased by 26% to 734.3 km^2 since the country's independence in 1965 [15, 16]. Over 60% of Singapore's coral reefs have been lost due to land reclamation and coastal armoring [14]. Large-scale coastal development activities have also increased sedimentation and turbidity levels that smother reef organisms and limit light penetration required for coral growth, consequently restricting their colonization and establishment to the shallows [17, 18].

Since the 1990s, at least five major coral relocation exercises have been carried out in Singapore as mitigation measures to counteract reef loss from areas designated for land reclamation [19]. In the early 2010s, plans were announced to consolidate port operations at the western coast to support Singapore's future needs, creating the mega container terminal, Tuas Megaport [20, 21]. An environmental impact assessment in 2012 determined that the planned dredging and reclamation activities would negatively impact the reef fringing Sultan Shoal, a small offshore island approximately 5 km southwest of the Singapore mainland [22, 23]. To mitigate the permanent loss of reef life from Sultan Shoal, environmental consultants relocated 2300 coral colonies to reefs at the southern offshore islands of St John's and Subar Darat [22, 23] (Fig. 10.1). Concurrently, the Maritime and Port Authority of Singapore and the Tropical Marine Science Institute also commenced research to maximize the conservation and restoration of scleractinian diversity. Making full use of 1380 loose coral fragments (or 'corals of opportunity') from 30 species, the overall aim of this research program was to build on existing reef restoration and rehabilitation strategies, improve the scientific understanding of processes necessary for the development of restored reefs, and design sustainable coastal management solutions. Here, we document the key lessons learned from this effort to enhance Singapore's coral reefs while maintaining its standing as a Green Port.

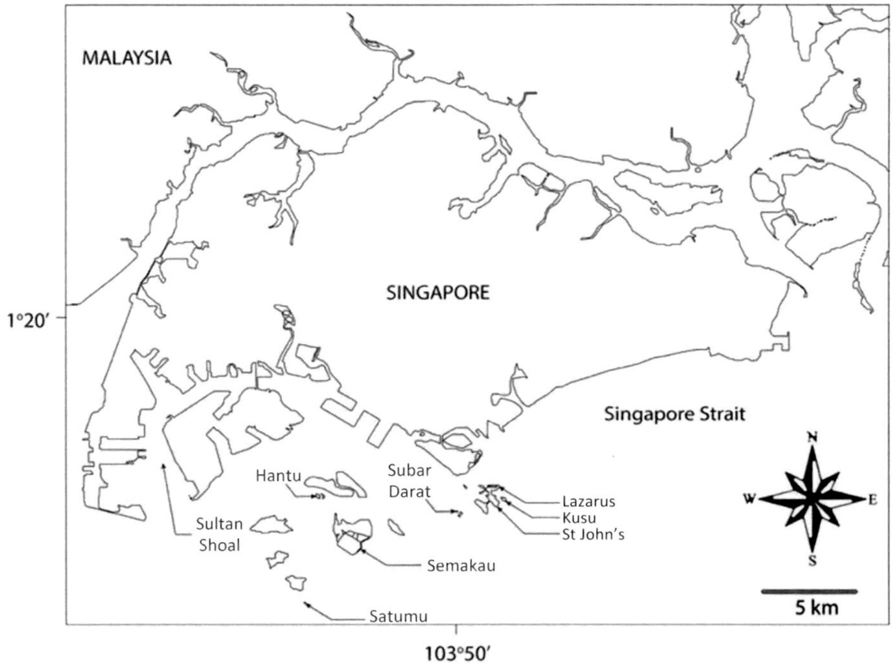

Figure 10.1 Map of Singapore and study sites.

10.2 Looking Before You Leap: Pre-Assessments Are Important for Reef Restoration

Identifying recipient sites with comparable and appropriate environmental conditions to donor sites will help minimize stress on coral propagules and augment their survival and growth rates [24–26]. To evaluate the suitability of recipient sites for Sultan Shoal's coral fragments, assessments were conducted in 2013 and 2014 at several locations. This included reefs fringing six of Singapore's southern offshore islands (Kusu, Hantu, Satumu, Semakau, St. John's, and Subar Darat) as well as two seawalls (Lazarus East and Lazarus West) (Fig. 10.1). Compared to Sultan Shoal, Semakau and Hantu registered higher sedimentation rates while Satumu and Subar Darat experienced much stronger currents [23]. These sites were thus considered less appropriate for receiving the coral fragments due to the greater risks of smothering by sediment or abrasion of soft tissues.

Based on similarities in environmental conditions (water quality, tidal range, site accessibility, and the presence of suitable substrates for transplantation) to Sultan Shoal, three sites were selected for the establishment of nurseries to rear coral fragments [23]. Lazarus East and Lazarus West were chosen as they are relatively sheltered environments with sandy seabeds that facilitated the installation of coral nursery frames. As coral communities have naturally established on many coastal defense structures in Singapore [27–29], there was also potential for the adjacent granite rock seawalls at both sites to be platforms for coral transplantation to create new reef habitats. In addition, the reef fringing northeastern Kusu Island was also identified as a rehabilitation site as it was a degraded environment with many dead coral boulders on which coral fragments could be transplanted.

Beyond site selection, the types and amount of corals to be relocated are also important considerations. Strategically selecting corals, by taking into account genetic diversity, can increase their resilience to environmental disturbances at the recipient site [30]. Our evaluation of four coral species (*Echinopora lamellosa*,

Merulina ampliata, Podabacia crustacea, and *Duncanopsammia peltata*) showed that approximately 40% of the total colony count represented 80% of the genetic diversity of each species at Sultan Shoal [31]. The analyses also showed that it was more beneficial to collect colonies of gonochoric (single-sex) species from a larger reef area to ensure sufficient genetic diversity. These findings represented a useful means of estimating the appropriate diversity of corals to be relocated from the donor site. Our use of 'corals of opportunity' (COPs) from across Sultan Shoal's **reef** also ensured that the relocated material was genetically varied.

10.3 Well-Designed Nurseries Improve Coral Recovery and Production

Coral nurseries, typically stocked with material fragmented from healthy parent colonies, help promote coral growth in a sheltered environment until the propagules attain a suitable size for transplantation [32–35]. They can also serve as repositories that preserve reef diversity against impending impacts from environmental changes such as coastal development, and at the same time generate new material for future restoration endeavors [36, 37]. Here, we focused on (1) optimizing nursery design and maintenance, and (2) increasing coral yield.

10.3.1 Nursery Design and Maintenance

In total, 98 nursery tables were deployed at Lazarus East, Lazarus West, and Kusu. To reduce sediment accumulation and facilitate the stable attachment of coral fragments, the nursery frames were elevated at least 0.5 m above the substrate and designed with mesh-net platforms (Fig. 10.2). We tested different designs and found that nursery tables with horizontal mesh platforms were more stable in areas with moderate wave action, compared to those with sloped or vertically oriented platforms. For fragments of species such as *Pocillopora acuta*, vertically oriented platforms were more detrimental to their survivorship [38]. In addition, macroalgae and biofouling organisms often colonized these nursery frames and inhibited coral growth. Monthly maintenance of the nurseries was necessary to reduce sediment accumulation and biofouling, and was effective at encouraging coral growth.

Figure 10.2 (A) Elevated nursery frames and (B) mesh net platforms helped reduce sediment accumulation and facilitated attachment of coral fragments.

10.3.2 Increasing Coral Yield

Instead of sourcing directly from healthy parent colonies (which could limit genetic diversity and introduce collateral damage), COPs which would have otherwise perished due to abrasion from currents and sediment were salvaged from Sultan Shoal's reef and progressively transferred to the coral nursery tables. Thirty species of COPs were collected over the course of the project, enabling a comprehensive evaluation of their responses in the nurseries. Overall survivorship after a year of rearing was high, at 92%. One year later, in May 2015, 15 species had high survivorship (>80%), while five species (*Merulina ampliata, Favites halicora, Favites pentagona, Porites rus,* and *Pectinia paeonia*) had survival rates of less than 50%. These observations helped to identify the species that fared the best in *in situ* nurseries under such conditions and shaped subsequent coral-rearing protocols in the project.

The coral nurseries also accelerated the recovery of fragmentation injuries and enhanced coral yield in Singapore's sedimented waters. Fragments reared in the nurseries grew significantly faster compared to those that were directly transplanted, although the benefit of a nursery phase on survival rates was species-specific [39]. However, while the nursery-reared corals of some species can be further fragmented to limit the need for harvesting from the wild, the growth rates of the new fragments may be reduced [14]. Such a method may need to be tested for its viability for different species.

10.4 Species Trait Characterization and Transplantation Protocol Customization Are Essential to Improve Restoration Outcomes

Coral transplantation is a common strategy to rapidly increase the amount of live coral cover at degraded reefs [25]. This approach may also be adapted to improve the biodiversity and ecological value of coastal defense structures [27]. Following a period of rearing, coral fragments were transplanted from the nurseries onto granite rock seawalls and degraded reefs at Lazarus and Kusu. This involved scrubbing the substrate thoroughly and using marine epoxy to attach the corals securely [27]. In addition to effectively promoting the colonization of reef biota at the study sites, our findings underscored the need to: (1) test and characterize species suitability for restoration, and (2) develop transplantation protocols for effective restoration outcomes.

10.4.1 Testing and Characterizing Species Suitability

In order to maximize the recovery of marine biodiversity and ecosystem functioning, it is necessary to test a variety of coral species for their suitability in restoration projects. Growth rates indicate suitability and allow for projections of transplant establishment at restoration sites to be made. However, information for only about 10% of the world's coral species is available in the literature [40], underscoring the need for more extensive and empirical determination of this key trait. While some species have registered high survival and growth rates from previous projects [39, 41, 42], understanding how a wider range of coral species fare within local contexts will increase long-term success rates. It is also essential to understand site- and taxon-specific responses to disturbance events (e.g., thermal stress), so that 'winners' and 'losers' can be identified to formulate more effective strategies [43].

Between 2015 and 2019, 1354 nursery-reared corals were progressively transplanted in batches to the seawalls and degraded reefs at Lazarus and Kusu (Fig. 10.3). These transplants contributed 7.9 m^2 of live coral area over four years. Targeted long-term monitoring of 163 transplants at Lazarus East revealed a high overall survivorship of 74.2%, but performances among species were wide-ranging. Transplants of three species (*Mycedium elephantotus, Pachyseris speciosa,* and *Turbinaria mesenterina)* did not survive, indicating their relative unsuitability as candidates for biodiversity restoration on seawalls. The surviving transplants of all other species grew larger, with the exception of *Echinophyllia aspera,* which registered a reduction in live tissue area (see Table 10.1; Figs. 10.4–10.6). Through our monitoring efforts, linear extension rates from transplants of 27 coral species were also determined, enabling projections of transplant establishment for future restoration efforts.

Figure 10.3 Coral fragments newly transplanted on a seawall at Lazarus Island, Singapore, in 2016.

Table 10.1 Survivorship and linear extension rates (mean ± S.D.) of corals transplanted on the Lazarus East seawall at Lazarus Island, Singapore

Species	Growth form[a]	Trans-plantation period (month)	Survivorship (%)	Linear extension rate (cm/year)
Acanthastrea echinata	Massive	37	100	1.3[b]
Acanthastrea rotundoflora	Encrusting	46	66.7	2.1 ± 0.2
Cyphastrea microphthalma	Massive	46	100	0.8[b]
Diploastrea heliopora	Massive	25	100	1.3 ± 0.2
Dipsastraea favus	Massive	25	90	1.3 ± 0.5
Duncanopsammia peltata	Laminar	46	50	1.2[b]
Echinophyllia aspera	Laminar	25	78.6	−0.02 ± 2.7[c]
Echinopora gemmacea	Laminar	37	100	3.0 ± 0.3
Echinopora horrida	Encrusting, long upright branches	41	14.3	2.2 ± 0.2
Favites complanata	Massive	37	100	0.8[b]
Favites halicora	Massive	46	100	1.6 ± 0.6
Favites pentagona	Sub-massive	46	50	1.2 ± 0.6
Goniastrea pectinata	Sub-massive	25	100	1.2 ± 0.3
Goniastrea retiformis	Massive	46	100	0.8[b]
Goniopora lobata	Columnar	46	100	1.8 ± 0.7
Hydnophora exesa	Sub-massive	37	50	4.3[b]
Leptoria phrygia	Massive	32	100	1.1 ± 0.5
Lithophyllon undulatum	Encrusting	37	100	0.4 ± 0.1
Lobophyllia hemprichii	Massive	37	50	1.0[b]
Lobophyllia recta	Massive	46	100	1.2 ± 0.1
Mycedium elephantotus	Laminar	13[d]	0	–

Species	Growth form[a]	Trans-plantation period (month)	Survivorship (%)	Linear extension rate (cm/year)
Pachyseris speciosa	Laminar	3[d]	0	–
Pavona cactus	Bifacial	32	72.7	2.0 ± 0.8
		31	80	1.7 ± 0.9
Pavona decussata	Bifacial	46	66.7	3.0 ± 0.1
Pavona explanulata	Laminar	32	87.5	0.6 ± 0.3
Pavona frondifera	Bifacial	37	100	2.4[b]
		31	33.3	2.9 ± 1.7
Porites rus	Digitate	25	70	1.5 ± 1.0
Psammocora contigua	Columnar	32	100	2.4 ± 0.5
		25	100	2.7 ± 0.4
Turbinaria mesenterina	Laminar	8[d]	0	–
Turbinaria stellulata	Massive	37	100	1.8 ± 0.1

[a]Growth forms from Ng et al. (2021) [8].

[b]Only one fragment transplanted or remaining.

[c]Transplants of *Echinophyllia aspera* decreased in size.

[d]Species that did not survive the duration of the study period (maximum period of survival indicated).

Figure 10.4 *Platgyra sinensis* (white arrow) transplants in 2018 (left), and other corals from the same batch of transplants in 2022 (right) (scale bars = 10 cm).

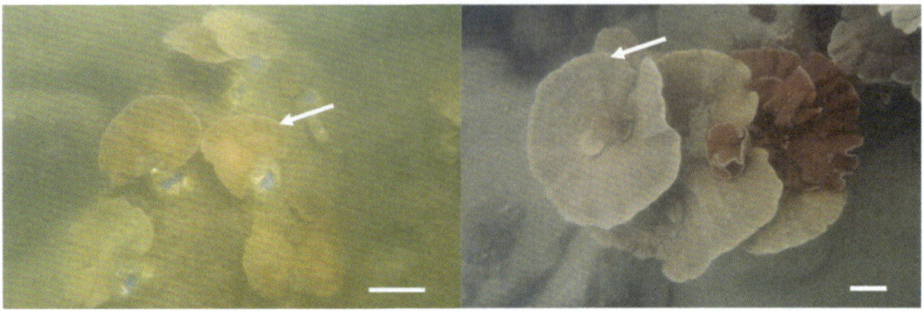

Figure 10.5 *Podabacia crustacea* (white arrow) transplants in 2016 (left), and other corals from the same batch of transplants in 2022 (right) (scale bars = 10 cm).

Figure 10.6 *Goniopora lobata* transplant in 2018 (left) and 2022 (right) (scale bars = 5 cm).

10.4.2 Developing Transplantation Protocols to Achieve Ecologically Effective Outcomes

While transplantation was an effective way of accelerating the colonization of scleractinians at the subtidal zone of seawalls [41], customizing existing protocols to suit local conditions further improved these outcomes. For example, transplantation carried out within periods that were less environmentally stressful, such as the inter-monsoon seasons contributed to increased coral growth. Additionally, trials at the Lazarus West seawall indicated that transplantation at shallower depths generated better yield (approximately 1.5 times faster growth rates for transplants at 3 m vs 6 m below chart datum) for certain species (e.g., *Pectinia paeonia* and *Turbinaria mesenterina*), likely due to different resource requirements. We also learned that transplantation in areas with dense macroalgal growth was ineffective, as the macroalgae can outcompete corals and adversely affect the survivorship and growth of coral transplants through allelochemical mechanisms or physical abrasion. In addition, Sam et al. (2021) [42] reported that although larger coral transplants (9–11 cm) have a higher chance of survival due to their reduced risk of complete mortality by predation and sedimentation burial, transplanting smaller coral fragments (2–4 cm) ultimately generated better yield in live coral cover and was the most efficient use of the original coral source material, thereby facilitating planning in the nursery-rearing phase.

10.5 Toward Long-Term Monitoring of Reef Restoration Efforts

Long-term coral monitoring is key to assessing whether transplants are able to establish at their recipient site, determining if ecosystem services can be provided by the established transplants, and understanding coral species' responses to threats such as corallivory and climate change. As a good indicator of their establishment at the recipient site, we observed that transplants of *Platygyra sinensis* (diameter 15–20 cm) on the Lazarus East seawall spawned naturally during a mass coral spawning event in April 2019 (Fig. 10.7). This indicated that, with proper monitoring and care, asexually propagated transplants can eventually develop into sexually mature colonies that can help seed and contribute genetic diversity to other locations.

Apart from tracking coral performance, it is just as important to document the recruitment of non-coral fauna over the longer term to comprehensively assess the overall contributions of the restoration effort to ecosystem functioning. Although *in situ* coral nurseries are used primarily to culture coral material [44], they can also provide shelter and food for reef-associated organisms. Marine animals from at least 100 taxa were documented at our *in situ* coral nurseries, underscoring their role as platforms for reef faunal recruitment (see Appendix; Fig. 10.8).

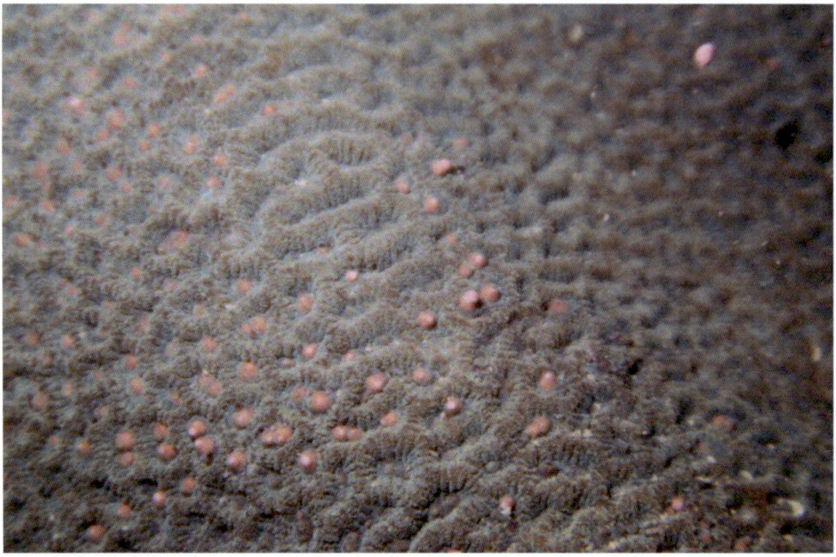

Figure 10.7 Three-year-old *Platygyra sinensis* transplant spawning in 2019.

Figure 10.8 Marine organisms utilizing the coral nurseries: (A) Cuttlefish (*Sepia* sp.), (B) Nudibranch eggs, (C) Spanish flag snapper (*Lutjanus carponotatus*), and (D) Spangled flatworm (*Acanthozoon* sp.).

Juvenile corallivorous butterflyfishes (*Chaetodon octofasciatus*) were observed swimming among and feeding on the *Acropora* spp. and *Pocillopora acuta* colonies that were reared in the nurseries [45]. Dense clusters of cuttlefish eggs were also deposited deep within coral branches, and juvenile cuttlefishes were seen seeking refuge around the *Acropora aculeus* and *P. acuta* colonies [46]. In addition, Wee et al. (2019) [47] reported that the nursery-reared corals also supported diverse epifaunal communities. From just 22 colonies of three coral species, 418 mobile invertebrates from 63 taxa such as crabs, shells, and shrimps were recorded. Each coral host species supported different epifaunal communities, likely as a consequence of different resource provisioning capacities. While a variety of epifauna also colonized dead coral colonies, these were of a different assemblage than those in live colonies.

Finally, coral monitoring regimes should also be complemented with regular checks of corallivores, which at high densities can cause mass coral mortality and negate restoration efforts. Snails, such as *Drupella spp.,* are known to feed exclusively on corals [48, 49]. They are nocturnal, preferring to hide under corals or boulders during the day and actively feed at night [50, 51]. As they also lay eggs on exposed coral skeletons [52], coral materials in nurseries and at transplant sites should be regularly examined and the presence of these snails should be recorded as part of coral monitoring protocols. The affected coral skeletons or egg capsules should be removed to reduce the chance of corallivore outbreaks.

10.6 Conclusion

Our research has highlighted how coral restoration efforts can be innovatively designed to mitigate the loss of coral cover and diversity from intensive coastal development, while also creating new reef communities in areas where reefs did not exist before. The success of site selection, coral rearing, and transplantation protocols as well as monitoring regimes was evident from the high coral survivorship rates and subsequent increases in marine biodiversity at the recipient sites. Sustained efforts in long-term monitoring are necessary to properly assess whether transplants are able to establish at their recipient site and to further determine if ecosystem services can be sustained by the established transplants.

Future Plans

While marine protected areas have an important role in the same way that forest reserves do, additional coral communities established via transplantation within Singapore waters will aid biodiversity conservation by providing "wildlife" corridors that provide ecological connectivity and additional refuges for marine life. If properly constructed, such reefs can also provide socio-economic value as they are a positive reflection of effective water quality management and vibrant ecosystem health in Singapore's Green Port. Concurrently, research should also be conducted into restoring a larger suite of ecosystem functions in degraded reefs or enhancing such functions in man-made habitats via the transplantation of other benthic groups [27]. It is equally important to consider the needs and perspectives of all stakeholders involved, so that reef restoration practitioners can create opportunities to build support and buy-in for the projects, increasing the likelihood of successful implementation and long-term success [53]. Toh et al. (2017) [41] calculated that the inclusion of trained volunteers in fieldwork (i.e., coral culture and transplantation) and data analyses was more cost-effective than if volunteers only assisted with fieldwork, or if only researchers were involved in the effort. Clearly, habitat conservation programs involving the community not only help reduce project expenses but also enhance capacity building and encourage a greater sense of public ownership of the natural heritage [41, 54].

Acknowledgments

This study is part of the project 'Enhancing Singapore's Coral Reef Ecosystem in a Green Port' supported by the Maritime and Port Authority of Singapore, under the project grant R-347-002-215-490.

Appendix

Reef fauna observed utilizing coral nurseries at Lazarus Island, Singapore

Phylum	Class	Order	Family	Genus	Species
Annelida	Polychaeta	Eunicida	Eunicidae	*Eunice*	*Eunice* sp.
				Marphysa	*Marphysa* sp.
				Nicidion	*Nicidion* sp.
		Phyllodocida	Chrysopetalidae		
			Iphionidae		
			Nereididae		
			Polynoidae		
			Syllidae		
			Terebellidae		
		Sabellida	Sabellidae		
		Scolecida	Capitellidae	*Capitella*	*Capitella* sp.
		Terebellida	Terebellidae		
Arthropoda	Hexanauplia	Copepoda			
	Malacostraca	Amphipoda	Ampeliscidae		
			Amphilochidae		
			Ampithoidae		
			Ischyroceridae		
			Leucothoidae		

Phylum	Class	Order	Family	Genus	Species
		Decapoda	Alpheidae	*Alpheus*	*Alpheus* sp.
				Athanas	*Athanas* sp.
				Synalpheus	*Synalpheus* sp.
			Diogenidae	*Dardanus*	*Dardanus lagopodes*
			Dromiidae	*Cryptodromia*	*Cryptodromia* sp.
			Galatheidae	*Galathea*	*Galathea coralliophilus*
			Galatheidae	*Galathea*	*Galathea johnsoni*
			Hippolytidae	*Hippolyte*	*Hippolyte ventricosa*
			Ischyroceridae		
			Palaemonidae	*Cuapetes*	*Cuapetes* sp.
				Phyllognathia	*Phyllognathia ceratophthalma*
			Pilumnidae	*Pilumnus*	*Pilumnus* sp.
			Porcellanidae	*Lissoporcellana*	*Lissoporcellana spinuligera*
					Pisidia streptochiroides
			Tetraliidae	*Tetralia*	*Tetralia nigrolineata*
			Trapeziidae	*Trapezia*	*Trapezia cymodoce*
				Chlorodiella	*Chlorodiella nigra*

Phylum	Class	Order	Family	Genus	Species
			Trapeziidae	*Cymo*	*Cymo andreossyi*
				Pilodius	*Pilodius cf. granulatus*
			Xanthidae	*Soliella*	*Soliella cf. flava*
		Isopoda	Sphaeromatidae		
		Mysida			
		Tanaidacea	Leptocheliidae		
	Ostracoda	Myodocopida	Cypridinidae		
Chordata	Actinopterygii	Gobiiformes	Gobiidae	*Gobiodon*	*Gobiodon histrio*
				Istigobius	*Istigobius ornatus*
		Perciformes	Apogonidae	*Ostorhinchus*	*Ostorhinchus margaritophorus*
			Blenniidae	*Petroscirtes*	*Petroscirtes variabilis*
			Carangidae	*Selaroides*	*Carangidae leptolepis*
			Chaetodontidae	*Chaetodon*	*Chaetodon octofasciatus*
				Chelmon	*Chelmon rostratus*
			Ephippidae	*Platax*	*Platax teira*
			Haemulidae	*Plectorhinchus*	*Plectorhinchus chaetodonoides*
					Plectorhinchus chrysotaenia
					Plectorhinchus gibbosus

Phylum	Class	Order	Family	Genus	Species
			Labridae	*Choerodon*	*Choerodon anchorago*
					Choerodon schoenleinii
				Halichoeres	*Halichoeres bicolor*
					Halichoeres leucurus
					Halichoeres nigrescens
				Scarus	*Scarus ghobban*
					Scarus rivulatus
			Latidae	*Psammoperca*	*Psammoperca waigiensis*
			Lutjanidae	*Lutjanus*	*Lutjanus carponotatus*
			Monacanthidae	*Acreichthys*	*Acreichthys tomentosus*
			Mullidae	*Upeneus*	*Upeneus tragula*
			Nemipteridae	*Pentapodus*	*Pentapodus paradiseus*
				Scolopsis	*Scolopsis bilineatus*
					Scolopsis monogramma
			Pinguipedidae	*Parapercis*	*Parapercis* sp.
			Pomacanthidae	*Chaetodontoplus*	*Chaetodontoplus mesoleucus*

Phylum	Class	Order	Family	Genus	Species
			Pomacanthidae	*Pomacanthus*	*Pomacanthus annularis*
					Pomacanthus sextriatus
				Dischistodus	*Dischistodus prosopotaenia*
				Neopomacentrus	*Neopomacentrus cyanomos*
					Neopomacentrus filamentosus
				Pomacentrus	*Pomacentrus cuneatus*
					Pomacentrus tripuncuatus
			Pseudochromidae	*Congrogadus*	*Congrogadus subducens*
			Serranidae	*Cephalopholis*	*Cephalopholis boenak*
				Diploprion	*Diploprion bifasciatum*
		Syngnathiformes	Syngnathidae	*Corythoichthys*	*Corythoichthys amplexus*
		Tetraodontiformes	Tetraodontidae	*Arothron*	*Arothron stellatus*
				Monacanthus	*Monacanthus chinensis*
Echinodermata	Holothuroidea	Synallactida	Stichopodidae	Stichopus	*Stichopus horrens*
	Ophiuroidea	Ophiurida	Ophiactidae	*Ophiactis*	*Ophiactis savignyi*
			Ophiotrichidae		

Phylum	Class	Order	Family	Genus	Species
Mollusca	Bivalvia	Arcida	Arcidae	*Barbatia*	*Barbatia amygdalumtostum*
				Mesocibota	*Mesocibota bistrigata*
		Mytilida	Mytilidae	*Musculus*	*Musculus* sp.
		Pteriida	Pteriidae		
	Cephalopoda	Cephalopoda	Octopoda	*Octopus*	*Octopus* sp.
		Sepiida	Sepiidae	*Sepia*	*Sepia* sp.
	Gastropoda	Caenogastropoda	Triphoridae	*Triphora*	*Triphora* sp.
		Littorinimorpha	Rissoidae	*Alvania*	*Alvania ogasawarana*
		Neogastropoda	Columbellidae	*Euplica*	*Euplica scripta*
				Mitrella	*Mitrella* sp.
				Zafra	*Zafra* sp.
			Muricidae	*Drupella*	*Drupella margariticola*
				Drupella	*Drupella rugosa*
		Nudibranchia	Chromodorididae	*Chromodoris*	*Chromodoris lineolata*
			Discodorididae	*Discodoris*	*Discodoris* sp.
			Dorididae	*Doriopsis*	*Doriopsis pecten*

Phylum	Class	Order	Family	Genus	Species
		Trochida	Trochidae	*Jujubinus*	*Jujubinus polychromus*
				Stomatolina	*Stomatolina rubra*
		Littorinimorpha	Cypraeidae	*Erronea*	*Erronea ovum*
	Polyplacophora				
Platyhelminthes	Rhabditophora	Polycladida	Pseudocerotidae	*Acanthozoon*	*Acanthozoon* sp.
				Nymphozoon	*Nymphozoon bayeri*
				Pseudobiceros	*Pseudobiceros hancockanus*
				Pseudoceros	*Pseudoceros concinnus*
					Pseudoceros bifurcus
					Pseudoceros indicus
					Pseudoceros laingensis

References

1. Souter, D., Planes, S., Wicquart, J., Logan, M., Obura, D., and Staub, F. (2021). Status of coral reefs of the world: 2020 report. Global Coral Reef Monitoring Network (GCRMN)/International Coral Reef Initiative (ICRI). Accessed 23 December 2023. https://gcrmn.net/2020-report/.
2. Burke, L., Reytar, K., Spalding, M., and Perry, A. (2011). Reefs at risk revisited. (World Resources Institute, Washington DC).
3. Firth, L. B., Knights, A. M., Bridger, D., Evans, A. J., Mieszkowska, N., Moore, P. J., O'Connor, N. E., Sheehan, E. V., Thompson, R. C., and Hawkins, S. J. (2016). Ocean sprawl: Challenges and opportunities for biodiversity management in a changing world. *Oceanography and Marine Biology: An Annual Review*, 54, pp. 189–262.
4. Lizcano-Sandoval, L. D., Londoño-Cruz, E., and Zapata, F. A. (2018). Growth and survival of Pocillopora damicornis (Scleractinia: Pocilloporidae) coral fragments and their potential for coral reef restoration in the Tropical Eastern Pacific. *Mar. Biol. Res.*, 14(8), pp. 887–897.
5. Burt, J. A., Al-Khalifa, K., Khalaf, E., AlShuwaikh, B., and Abdulwahab, A. (2013). The continuing decline of coral reefs in Bahrain. *Mar. Pollut. Bull.*, 72(2), pp. 357–363.
6. Hughes, T. P., Huang, H., and Young, M. A. (2013). The wicked problem of China's disappearing coral reefs. *Conservation Biology*, 27(2), pp. 261–269.
7. Bozec, Y. M., Acosta-González, G., Núñez-Lara, E., and Arias-González, J. E. (2008, July). Impacts of coastal development on ecosystem structure and function of Yucatan coral reefs, Mexico. *Proceedings of the 11th International Coral Reef Symposium*, 18(18), pp. 691–695.
8. Ng, C. S. L., Chan, Y. K. S., Nguyen, N. T. H., Kikuzawa, Y. P., Sam, S. Q., Toh, T. C., Mock, A. Y. J., Chou, L. M., and Huang, D. (2021). Coral community composition and carbonate production in an urbanized seascape. *Marine Environmental Research*, 168, 105322.
9. Gayle, P. M., Wilson-Kelly, P., and Green, S. (2005). Transplantation of benthic species to mitigate impacts of coastal development in Jamaica. *Revista de Biología Tropical*, 53, pp. 105–115.
10. Kotb, M. M. (2016). Coral translocation and farming as mitigation and conservation measures for coastal development in the Red Sea: Aqaba case study, Jordan. *Environmental Earth Sciences*, 75, pp. 1–8.

11. Ku'ulei, S. R., Donà, A. R., Stender, Y., Lager, C., and Jokiel, P. L. (2017). Effectiveness of coral relocation as a mitigation strategy in Kāne 'ohe Bay, Hawai 'i. *PeerJ*, 5, e3346.
12. Liñán-Rico, V., Cruz-Ramírez, A., Michel-Morfín, J. E., and Liñán-Cabello, M. A. (2020). Rescue and relocation of benthic organisms during an urban-port development project: Port of Manzanillo, case study. *Aquat. Conserv. Mar. Freshwater Ecosyst.*, 30(6), pp. 1137–1148.
13. Sautya, S., Gaikwad, S., De, K., Nanajkar, M., Pradhan, U. K., Mohandass, C., and Sahu, B. (2022). Translocation of intertidal corals in highly urbanized Mumbai seascape: A pragmatic management initiative towards ecosystem conservation. *Ecol. Eng.*, 180, 106666.
14. Chou, L. M., Huang, D., Tan, K. S., Toh, T. C., Goh, B. P. L., and Tun, K. (2019). Chapter 24: Singapore. ed. Sheppard, C., *World Seas: An Environmental Evaluation* (Second Edition) Volume II: The Indian Ocean to the Pacific (Academic Press), pp. 539–558.
15. Tun, K. P. P. (2012). Optimisation of reef survey methods and application of reef metrics and biocriteria for the monitoring of sediment-impacted reefs. Dissertation, National University of Singapore.
16. Singapore Department of Statistics (2023). Environment. Accessed 2 April 2023. https://www.singstat.gov.sg/publications/reference/ebook/society/environment.
17. Erftemeijer, P. L., Riegl, B., Hoeksema, B. W., and Todd, P. A. (2012). Environmental impacts of dredging and other sediment disturbances on corals: A review. *Mar. Pollut. Bull.*, 64(9), pp. 1737–1765.
18. Ng, C. S. L., Toh, T. C., and Chou, L. M. (2013). Current status of coral reef restoration in Singapore. In *Proceedings of the Asian Conference on Sustainability, Energy & the Environment*, pp. 546–558.
19. Ng, C. S. L., and Chou, L. M. (2018). Coral reef restoration in Singapore – Past, present and future. In *Sustainability Matters: Environmental Management in the Anthropocene*, pp. 3–23.
20. Lam, J. S. L. (2016). Strategy of a transhipment hub: The case of Port of Singapore. In Lee, T. W and Cullinane, K. (Eds.). Dynamic Shipping and Port Development in the Globalized Economy. *Applying Theory to Practice in Maritime Logistics*, vol. 1, Palgrave Macmillan UK, pp. 12–38.
21. Ministry of Transport (2022). Enhancing Singapore's Connectivity; Securing Our Future. Press release. Accessed 2 April 2023. https://www.mot.gov.sg/news/press-releases/Details/enhancing-singapore-s-connectivity-securing-our-future.

22. Maritime and Port Authority (2015). Relocated corals thriving at new sites. Accessed 2 April 2023. https://www.mpa.gov.sg/media-centre/details/relocated-corals-thriving-at-new-sites.

23. Chou, L. M., Toh, T. C., and Ng, C. S. L. (2017). Effectiveness of reef restoration in Singapore's rapidly urbanizing coastal environment. *International Journal of Environmental Science and Development*, 8(8), pp. 576–580.

24. Chou, L. M., Yeemin, T., Yaman, A. R. B. G., Vo, S. T., and Alino, P. (2009). Coral reef restoration in the South China Sea. *Galaxea, Journal of Coral Reef Studies*, 11(2), pp. 67–74.

25. Edwards, A. J. (2010). Reef Rehabilitation Manual. (Coral Reef Targeted Research & Capacity Building for Management Program, St Lucia, Australia).

26. Bongiorni, L., Giovanelli, D., Rinkevich, B., Pusceddu, A., Chou, L. M., and Danovaro, R. (2011). First step in the restoration of a highly degraded coral reef (Singapore) by *in situ* coral intensive farming. *Aquaculture*, 322, pp. 191–200.

27. Ng, C. S. L., Lim, S. C., Ong, J. Y., Teo, L. M. S., Chou, L. M., Chua, K. E., and Tan, K. S. (2015). Enhancing the biodiversity of coastal defence structures: Transplantation of nursery-reared reef biota onto intertidal seawalls. *Ecol. Eng.*, 82, pp. 480–486.

28. Chou, L. M., Sam, S. Q., Toh, T. C., and Ng, C. S. L. (2018). The relevance of reef restoration in the continually changing marine environment of Singapore. *Vietnam Journal of Marine Science and Technology*, 18(3), pp. 278–285.

29. Kikuzawa, Y. P., Ng, C. S. L., Toh, T. C., Sam, S. Q., Lee, Y. L., Loo, P. L., Chua, Y. Z, Tan, K. S., and Chou, L. M. (2020). Diversity of subtidal benthic and hard coral communities on sloping and vertical seawalls in Singapore. *Marine Biodiversity*, 50, pp. 1–14.

30. Baums, I. B., Baker, A. C., Davies, S. W., Grottoli, A. G., Kenkel, C. D., Kitchen, S. A., Kuffner, I. B., LaJeunesse, T. C., Matz, M. V., Miller, M. W., Parkinson, J. E., and Shantz, A. A. (2019). Considerations for maximizing the adaptive potential of restored coral populations in the western Atlantic. *Ecol. Appl.*, 29(8), e01978.

31. Afiq-Rosli, L., Huang, D., Toh, T. C., Taira, D., Ng, C. S. L., Song, T., and Chou, L. M. (2019). Maximising genetic diversity during coral transplantation from a highly impacted source reef. *Conserv. Genet.*, 20, pp. 629–637.

32. Epstein, N., Bak, R. P. M., and Rinkevich, B. (2003). Applying forest restoration principles to coral reef rehabilitation. *Restor. Ecol.*, 9(4), pp. 432–442.
33. Shafir, S., and Rinkevich, B. (2010). Integrated long-term mid-water coral nurseries: A management instrument evolving into a floating ecosystem. *University of Mauritius Research Journal*, 16, pp. 365–386.
34. Rinkevich, B. (2005). Conservation of coral reefs through active restoration measures: Recent approaches and last decade progress. *Environ. Sci. Technol.* 39, pp. 4333–4342.
35. Ng, C. S. L., and Chou, L. M. (2014). Rearing juvenile 'corals of opportunity' in *in situ* nurseries–A reef rehabilitation approach for sediment-impacted environments. *Mar. Biol. Res.* 10(8), pp. 833–838.
36. Shafir, S., and Rinkevich, B. (2010). Integrated long-term mid-water coral nurseries: A management instrument evolving into a floating ecosystem. *University of Mauritius Research Journal*, 16, pp. 365–386.
37. Schopmeyer, S. A., Lirman, D., Bartels, E., Byrne, J., Gilliam, D. S., Hunt, J., Johnson, M. E., Larson, E. A., Maxwell, L., Nedimyer, K., and Walter, C. (2012). *In situ* coral nurseries serve as genetic repositories for coral reef restoration after an extreme cold-water event. *Restor. Ecol.*, 20(6), pp. 696–703.
38. Poquita-Du, R. C., Toh, K. B., Toh, T. C., Ng, C. S. L., Taira, D., Loke, H. X., Afiq-Rosli, L., Chou, L. M., Song, T. C., and Cabaitan, P. (2017). Effects of nursery table slope orientation on coral survival and growth. *Mar. Biol. Res.*, 13(9), pp. 975–982.
39. Afiq-Rosli, L., Taira, D., Loke, H. X., Toh, T. C., Toh, K. B., Ng, C. S. L., Cabaitan, P. C., Chou, L. M., and Song, T. (2017). *In situ* nurseries enhance coral transplant growth in sedimented waters. *Mar. Biol. Res.* 13(8), pp. 878–887.
40. Madin, J. S., Hoogenboom, M. O., Connolly, S. R., Darling, E. S., Falster, D. S., Huang, D., Keith, S., Mizerek, T., Pandolfi, J. M., and Baird, A. H. (2016). A trait-based approach to advance coral reef science. *Trends. Ecol. Evol.*, 31(6), pp. 419–428.
41. Toh, T. C., Ng, C. S. L., Loke, H. X., Taira, D., Toh, K. B., Afiq-Rosli, L., Du, R. C. P., Cabaitan, P., Sam, S. Q., Kikuzawa, Y. P., and Chou, L. M. (2017). A cost-effective approach to enhance scleractinian diversity on artificial shorelines. *Ecol. Eng.*, 99, pp. 349–357.
42. Sam, S. Q., Ng, C. S. L., Kikuzawa, Y. P., Toh, T. C., Sim, W. T., and Chou, L. M. (2021). Influence of fragment size on post transplantation

growth and survival of domed scleractinian corals. *Mar. Biol. Res.*, 17(4), pp. 327–340.

43. Ng, C. S. L., Huang, D., Toh, K. B., Sam, S. Q., Kikuzawa, Y. P., Toh, T. C., Taira, D., Chan, Y. K. S., Hung, L. Z. T., Sim, W. T., Rashid, A. R., Afiq-Rosli, L., Ng, N. K., and Chou, L. M. (2020). Responses of urban reef corals during the 2016 mass bleaching event. *Mar. Pollut. Bull.*, 154, 111111.
44. Rinkevich, B. (2014). Rebuilding coral reefs: Does active reef restoration lead to sustainable reefs? *Current Opinion in Environmental Sustainability*, 7, pp. 28–36.
45. Taira, D., Toh, T. C., Ng, C. S. L., Sam, S. Q., and Chou, L. M. (2016). Coral nurseries as habitats for juvenile corallivorous butterflyfish. *Marine Biodiversity.*, 47(3), pp. 787–788.
46. Sam, S. Q., Taira, D., Ng, C. S. L., Toh, T. C., and Chou, L. M. (2018). First record of cuttlefish egg-deposition at an *in situ* coral nursery. *Galaxea, Journal of Coral Reef Studies*, 20(1), pp. 9–10.
47. Wee, S. Y. C., Sam, S. Q., Sim, W. T., Ng, C. S. L., Taira, D., Afiq-Rosli, L., Kikuzawa, Y. P., Toh, T. C., and Chou, L. M. (2019). The role of *in situ* coral nurseries in supporting mobile invertebrate epifauna. *J. Nat. Conserv.*, 50, 125710.
48. Taylor, J. D. (1976). Habitats, abundance and diets of muricacean gastropods at Aldabra Atoll. *Zool. J. Linn. Soc.*, 59(2), pp. 155–193.
49. Cumming, R. L., and McCorry, D. (1998). Corallivorous gastropods in Hong Kong. *Coral Reefs*, 17(2), pp. 178–178.
50. Moyer, J. T. (1982). Massive destruction of scleractinian corals by the muricid gastropod, Drupella, in Japan and the Philippines. *The Nautilus*, 96, pp. 69–82.
51. Boucher, L. M. (1986). Coral predation by muricid gastropods of the genus Drupella at Enewetak, Marshall Islands. *Bull. Mar. Sci.*, 38(1), pp. 9–11.
52. Sam, S. Q., Toh, T. C., Kikuzawa, Y. P., Ng, C. S. L., Taira, D., Afiq-Rosli, L., Tan, K. S., and Chou, L. M. (2017). Egg capsules and veligers of the corallivorous muricid gastropod Drupella rugosa (Born, 1778). *Invertebr. Reprod. Dev.*, 61(3), pp. 164–171.
53. Ng, C. S. L., Toh, T. C., Toh, K. B., Sam, S. Q., Kikuzawa, Y. P., Chou, L. M., and Huang, D. (2022). Enhancing reef restoration by assessing stakeholder knowledge, attitudes and preferences. *Restor. Ecol.*, e13854.
54. Low, J., and Lim, L. J. (2012). Conservation of our natural heritage: The Singapore experience. In *Proceedings of the 12th International Coral Reef Symposium, Cairns*, Australia, pp. 10–13.

Index

For Product Safety Concerns and Information please contact our EU representative GPSR@taylorandfrancis.com Taylor & Francis Verlag GmbH, Kaufingerstraße 24, 80331 München, Germany